ZERO WASTE-SUSTAINABLE ENVIRONMENT: GREEN ENVIRONMENT AND SOCIAL SIGNIFICANCE

Other Related Books

Title	*Author*
A Handbook of Environment Impact Assessment	*V.S. Kulkarni*
Antarctica: Environment and Conservation (Arising Issues, Policies and Management)	*Neloy Khare*
Biodiversity Conservation	*M.S.B. Kumar*
Biodiversity Conservation and Systematics	*Prithipal Singh*
Biodiversity Endangered	*A.B. Chaudhuri*
Biological Sciences : A Kaleidoscopic View of New Trends	*R.P. Saran*
Climate Change and Environment	*J. Sundaresan*
Climate Change and Sustainable Development	*Om Prakash*
Climate Change Impact on Ecosystem	*J. Sundaresan*
Climate, Agriculture and Man (Shrinking Biodiversity and Sustainability)	*V.K. Sehgal*
Combating Desertification in Developing Countries	*J.K. Jain*
Consciousness, Entropy and EVolution: The Laws of Biology	*DMR Sekhar*
Disaster : Perception and Management	*T.S. Chouhan*
Drinking Water: Quality Assessment and Management	*S.K. Gupta*
Eco-Degradation Due to Air Pollution	*Arun Arya*
Ecology Workbook	*R. Misra*
Environment and Self-Endangered Man	*J.V. Amin*
Environment at Crossroads Challenges and Green Solutions	*Arun Arya*
Environmental Pollution and Environmental Management	*P. Dwivedi*
Essentials of Environmental Studies	*S.S. Sagwal*
Genopsych: A Coinage in The Foundry of Biology	*DMR Sekhar*
Greenhouse Engineering	*R.A. Aldrich*
Industrial Pollution and Environmental Management	*R.K. Trivedy*
Industrial Waste Waters and Environmental Pollution	*D.C. Joshi*
Land Utilisation and Resource Management	*Lalit Singh Jhala*
Microplastic Pollution and Risks	*Golam Kibria*
Pollution Risks, Impacts & Management Social, Economic and Environmental Perspectives	*Golam Kibria*
Rainwater Management Theory and Practice	*M.L. Jat*
Urban Solid Waste Management in India	*U.S. Sree Ramulu*
Wasteland Management and Environment	*A.K. Roy*
Water Analysis Guide	*HACH*
Water Pollution and Quality Management (Theory and Practicals)	*S. K. Gupta*

ZERO WASTE-SUSTAINABLE ENVIRONMENT: GREEN ENVIRONMENT AND SOCIAL SIGNIFICANCE

Dr. Jyoti Prakash[1], Dr. Rachna Chaturvedi[1], Dr. Ruchi Yadav[1], Ajai Prakash[2]

1. Amity Institute of Biotechnology, Amity University Uttar Pradesh, Lucknow, Uttar Pradesh, India
2. Department of Business Administration, University of Lucknow, Lucknow, Uttar Pradesh, India

SCIENTIFIC® PUBLISHERS
New Delhi • Jodhpur

SCIENTIFIC® PUBLISHERS

Published by:

Jodhpur
5 A, New Pali Road, P.O. Box 91
Jodhpur - 342 001 Rajasthan (India)
E-mail: info@scientificpub.com
Website:
www.scientificpubonline.com

New Delhi
4806/24, Ansari Road, Daryaganj,
New Delhi - 110 002 (India)

ISBN: 9788196246440
eISBN: 9788196246921

Printed in India

PREFACE

Today's consumption economy brings its own challenges and waste disposal is an essential concern globally. It requires the attention of various stakeholders especially researchers and academicians who are looked up to provide solutions of this problem. This compilation has been developed keeping all aspects of Zero Waste for a sustainable environment. The present book will support to provide proposals to implement the best feasible approaches at ground levels and exhibit how science and technology, law and policies, modern societal procedures and indigenous knowledge are developing sustainable solutions, with the objective of zero waste. Zero waste shapes our world by embedding the concept of sustainability looking at environmental, economic, and social aspects for complete sustainable development.

There are a total of 09 chapters in this book. Chapter one is enlightened about the knowledge of waste management which will be instrumental in developing a sustainable solution through grass-root initiatives. In chapter two roles of microbes are discussed by using different techniques for a sustainable and green environment. Chapter three emphasizes the role of social media in handling issues of zero waste. Chapter four analyses recent advancements toward recycling processes of waste for sustainable waste management systems. Chapter five looks at an important eco-friendly tool that works as phytoremediation for wastes related to specific industries. Chapter six looks at the legal aspects and policies of India for addressing sustainable waste management. Chapters seven and eight share aspects related to computer waste management for zero waste structure. They look at environmental informatics which builds on current research and future prospects. The authors have discussed the recent scope of bioinformatics for the sustainable development of the environment. In Chapter nine different methods and challenges faced during the execution of tasks related to the journey of management of biomedical waste to zero waste have been treaded. As biomedical waste is a big concern for public health this chapter exposes the readers about the precaution to be taken for biomedical waste disposal.

This book will be beneficial to researchers, academicians, students, engineers and industry to work toward zero waste and a sustainable environment.

EDITORS BIOGRAPHY

Dr. Jyoti Prakash is an Asst. Prof. at Amity Institute of Biotechnology, Amity University Uttar Pradesh, Lucknow India. She has completed her M.Sc. and Ph.D. in Environmental Sciences and specialization in Air pollution. She has also served as a visiting faculty with United States International University in Nairobi, Kenya. She has published a number of research papers with high impact factors in reputed journals. She has also published edited book and various chapters. She is a regular contributor to television and media shows on environment and allied issues.

Dr. Rachna Chaturvedi is an Asst. Prof. at Amity Institute of Biotechnology, Amity University Uttar Pradesh, Lucknow India. Recipient of Chancellor's gold medal at M.Sc. level. Obtained PhD from RML University Faizabad. Dedicated research efforts in microbiology for last two decades. Published many research papers in international and national journals. She has also published edited book and various chapters. Appointed as Editorial Board members and reviewer in several international and national journals. Presented over 75 papers in international and national conferences. An active life member of many reputed societies.

Dr. Ruchi Yadav is an Asst. Prof. at Amity Institute of Biotechnology, Amity University Uttar Pradesh, Lucknow India. She has completed her M.Tech. and Ph.D. in Biotechnology with specialization in bioinformatics. Has 13 years of experience in teaching and research in area of Bioinformatics and computational biology. Has published several research papers in national and international journals of high repute and she has also presented and published more than 80 abstracts in symposia, national and international conferences. She has been awarded with patent and has keen interest in solving biological and environmental problems with the help of bioinformatics and computational biology approach.

Professor Ajai Prakash is a faculty in the Department of Business Administration University of Lucknow India. He has worked with United Nations on sustainable development issues and is a member of responsible management education wing which has been instituted to foster where issues of Sustainable development among students is. He has been graduate school at KCA University Nairobi Kenya. He is extensively engaged in issues related to sustainable development. He has contributed books and research articles in the reputed journals.

CONTENTS

Chapter 1

Embedding indigenous knowledge in waste management: Promoting sustainability through grass root innovation

Ajitesh Tiwari[1], Ajai Prakash[1] and Jyoti Prakash[2*]

[1] Department of Business Administration, University of Lucknow, Lucknow, Uttar Pradesh, India

[2] Amity Institute of Biotechnology, Amity University Uttar Pradesh, Lucknow, Uttar Pradesh, India

Abstract

The piling waste stock of India presents an immediate concern for communities, policy planners and implementers to address the glaring environmental, and public health concerns. The current phase of waste management technology development is fraught with challenges of cost affordability and scale of applications. The imperative to look for traditional wisdom to ameliorate the situation is promising and emanates from the fact that indigenous knowledge is rooted in the ethos of the land and hence easily adoptable. This piece of research analyses different indigenous practices for channelization of a variety of waste generated in agricultural, domestic and local industrial operations including grey water management. It further goes to probe the practices materializing the conception of waste to wealth.

Keywords: Indigenous knowledge, waste management, sustainability

* Corresponding Author's Email: jprakash@lko.amity.edu

Introduction

The ancient Indian knowledge imbibed the idea of matter in the scheme of its eternity of existence. The Vaisheshika school of Indian philosophy conceptualized the idea of atom well before the western philosophers deliberated upon it. It was also enshrined in this system of philosophy that matter cannot be destroyed; only forms can change and energy can transfer. The very idea of indestructibility of matter brought the notion of perennial utility of all matter encompassing human habitat (Narayan, 2002). In the given context, there is a historical tradition of sustainable waste management in India, in terms of devising ways of multiple life cycles of matter. Such knowledge systems although still in practice in different spheres and geographical locations but remain confined to a local area and now are becoming a part of the discussion of sustainable waste management. Herein, it is pertinent to discuss a few important practices which offer valuable insights to design a fusion of policies for sustainable waste management.

For thousands of years, animal husbandry remains a prime occupation of societies inhabiting Indian subcontinent. Ingenious people of the ancient era developed a practice of using the animal wastage to fuel source, devising a new life cycle using residual energy of waste generated by animals remains source of fuel due to various harms to human surroundings and health conditions, however, imagine the period when it would have begun to be used a source of fuel. Moreover, animal waste still forms the primary source of organic manure, for chemical fertilizer free agriculture which has become a subject of debate because of the repercussion soil losing its nutrients and groundwater contamination in the areas where intensive industrially produced fertilizer is growing in use for the last few decades.

In the rural hinterland of the Indian subcontinent we still come across houses made with mud and straw as well as by other materials locally available, such houses, when they come out of present life cycle, the material finds use in other construction activities.

Innovation and entrepreneurship in bio energy and biomass sector

The biogas is obtained from cow dung and other animal wastes, it can be harnessed for various utilities such as cooking, heating, generation of electricity and lighting purposes. Therefore it offers an immense potential to our rural masses to fulfill their energy needs about. About 5,000,000 family type biogas plants have been installed and national biogas and manure management program which is a flagship scheme of ministry of new and renewable energy, Government of India. The enormous potential of biomass in terms of its energy utility has been rightly identified by various state governments across the country. States like Haryana are offering a program to farmer's wherein they can supplement their earnings by offering the animal wastes such as cow dung another animal waste. The "GOBAR DHAN" scheme pronounced by the Haryana government offers a unique solution using traditional wisdom of harnessing energy potential (Government of Haryana 2018).

This scheme is offering both a supplementary income to farmers and promoting a culture of utilizing organic fertilizers for organic farming and reducing the amount of chemical fertilizers used which are made by industrial production process. Fertilizers produced this way have both environmental impacts in terms of reduced emission of greenhouse gases and reduction in emission of other environmental pollutants. On the other hand, industrially produced chemical fertilizers affect the soil fertility in an adverse manner in the long run (Jiai Liu, 2021). The growing soil salinity of the erstwhile green revolution areas such as Punjab and Haryana is an apt example of the adverse impact of the use of chemical fertilizers such as urea and nitrogen for a prolonged period in the fields (Kang,2021). The "GOBAR DHAN" schemes and other such schemes using bio waste and harnessing the potential of bio waste in making organic manure utilize the traditional wisdom continued over centuries in India and cutting both the adverse environmental impacts of the soil salinity and water pollution. Additionally, they utilize the waste in a way to create wealth and also they prevent the use of such biomass energy in the erstwhile ways which used to generate pollution through dust and ashes. Thus, the success of "GOBAR DHAN" scheme and other such schemes would establish the integral place of traditional wisdom in environmental conservation in a sustainable way.

Entrepreneurship ideas based on biomass

Biomass obtained from plants and trees including a variety of other organic materials such as paddy residue, sugar cane residue, straw, shells of coconut etc. offer huge potential in terms of generation of electricity and energy. The concept of waste to energy has been recognized duly by the policymakers and entrepreneurs of our country. And as evidence of this we note that more than 10 megawatts of electric capacity have been installed which utilizes biomass as the feed material. The evolving technology of generating power from biomass promotes the efficient use of such waste resources and provides a unique way to satisfy the growing energy needs of a burgeoning population of the country (Surie 2017).

The national policy on bio-fuels 2018 and the entrepreneurship opportunities in bio-fuel sector

Recognizing the huge outgo of foreign exchange and mounting current account deficit due to the large imports of fossil fuels such as diesel, petrol, aviation turbine fuel etc. from foreign countries; India has recognized the prime place of harnessing the mixed fuel technology and producing bio-fuels to mix with the fossil fuels. Traditionally some specified plants like jatropha had been recognized by our scientists for their potential to be offered as alternative to fossil fuels. Whereas relying on traditional wisdom the science has also recognized the methods which could be offered by bio-fuels to efficiently meet the energy needs of the country. The bio-fuels can be blended which fossil fuels such as petrol diesel and aviation turbine fuel and could reduce the import needs of the country by leaps and bounds. The new policy enunciated by Government of India in 2018 can offer a multitude of opportunities for farmers, processors, aggregators, transporters, entrepreneurs in the arena of bio-fuel supply chain to differentiate between the sources of bio fuels in terms of their efficiencies the policy categorize is the bio-fuels into three categories.

First generation biofuels

These biofuels are obtained from plants animal waste, food waste substances containing high amounts of sugar, starch, animal fats and oils from the vegetables. They are also obtained by using cereal crops and other commercial crops. The production of these biofuels involves sufficient application of labour and other resources.

Second generation biofuels

These are called advanced bio-fuels or fuels of the future; for they involve utilization of non food biomass such as waste materials of crops, plant waste such as wooden chips other substances which have the residual energy in the form of hydrocarbons. The discerning feature of these bio-fuels lies in the fact that they use residual agricultural material and waste plant produce to produce energy fuels and thus save both the precious earth resources and provide a source of reliable income to farmers and other stakeholders included in the bio-fuel supply chain.

These fuels contribute to management of environment by offering two-way positive feedback mechanism. On the one hand they reduce the consumption of fossil fuels; on the other hand they reduce d burning of waste materials in the fields and increase in the air pollution. The same idea is the motivation behind introducing many plant residue management schemes introduced in Delhi-NCR by the Government of India. These schemes offer benefits to farmers while managing environment and providing raw material for generation of second generation bio-fuels.

Third generation bio-fuels

Enormous potential of energy generation is offered by harnessing algae in a scientific manner and occupy very less area when compared to effort and area required under other methods of same level of bio-fuels by the second generation bio-fuels sources (Tripathi 2022). Third generation bio-fuels find a varied used destinations from bio diesels to jet fuel. Therefore the advanced bio-fuels have a potential to catapult the Indian economy to a carbon efficient economy with establishing an attractive entrepreneurship opportunity for farmer's, aggregators, processors, transporters and other value chain partners in the bio-fuel manufacturing supply chain.

Realizing the potential of advanced bio-fuels the government has come up with an enabling mechanism in this policy by allowing to arrange source materials such as sugar beet, sorghum, corn, cassava, damaged food grains rotten vegetables and other input such as molasses for ethanol production and other agricultural as well as industrial processes. The government is incentivizing the advanced bio-fuels through measures such as tax benefits assured purchase price and subsidies to the refineries to the tune of 5000 crores in a period of 6 years.

The policy underlying the notion of using waste to produce energy and promoting self reliance in terms of energy needs has far reaching consequences. Given the target of 20% blending of ethanol in petrol and 5% of biodiesel in diesel by 2030 billions of dollars of foreign reserves will be saved by these initiatives and about 1,000,000 jobs may get created in the entire supply chain of bio fuels. This will be accomplished by utilizing the traditional bio resources in an innovative manner offers huge opportunity for national entrepreneurial and environmental purposes.

Opportunities for small and medium entrepreneurship using traditional materials and age old wisdom

Traditional materials and methods off commercializing certain products made from biodegradable inputs have the capacity to offer recyclable, environmentally sustainable and economically viable products offering value for money to the customer and generating a huge number of employments in decentralized industries. For example the following ubiquitous products may be taken up at micro, small and medium industries and offered to the market with environment friendly tagline.

- Banana leaf plates which are enduring as well as biodegradable instead of thermocol plates which have been banned now.

- Bamboo based furniture instead of plastic furniture. Bamboo as a material offers durability has a comparatively prolonged life cycle vis-à-vis certain plastic based furniture.
- Utilizing water phased out from reverse osmosis process for micro irrigation for vegetables in the kitchen garden.
- Utilizing plastic waste in construction and recreation activities to ensure longevity of structures built, as plastic does not degrade by itself.
- Decentralized and localized vermin composting in urban areas to supplement the need for organic manure in rural and peri-urban areas.
- Developing aesthetic habits of turning waste into decorative items which may help managing non biodegradable waste in a manner that does not harm the environment.

Waste to wealth concept

The age-old wisdom prevailing in India that everything has a value is the idea behind the Government of India's mission. The materials discarded from one process as waste may a value somewhere else and so what is required at the end of the policy makers and entrepreneurs is to identify the post current life cycle uses of the waste produce (Boojh 2022). Working on these principles the government of India has launched waste to wealth mission to create an economy that ensures high amount of recycling, reusing and reintroducing the same materials across different value chains to realize the potential of circular economy (Government of India n.d.).

Zero effect zero defect scheme

In order to minimize resource wastage and mitigate pollution emanating from manufacturing and processes in industries, the Ministry of Micro, Small and Medium Enterprises (MSME) has come up with an initiative to introduce both qualitative and quantitative improvements in waste minimization and waste management. The scheme incentivizes industries to produce quality products that have no defect and simultaneously reduce the wastage generated from such production process. The scheme may attain huge success in removing green house gas emissions and prolonging material use cycle to effect net positive outcome on environment and employment (International Energy Agency 2021).

Initiatives needed for augmenting application of traditional knowledge in sustainable waste management

- A national mission on sustainable waste management technologies utilizing traditional sources and methods shall be launched to gain an in depth knowledge of the various sustainable waste management systems prevalent in Indian society from ancient to recent times.
- A behavior change campaign shall be launched at nationwide level to sensitize and conscientization of people towards the utility of simple traditional best practices in promoting sustainable waste management industry.
- Launching nationwide hackathons on traditional waste management practices imbibing sustainability shall be launched to crowd source ideas and bring out the methods being used using traditional knowledge in sustainable waste management.
- Diffusion of traditional waste management practices and modern technologies may help to catapult waste management in a sustainable manner; for this startups and entrepreneurs must be encouraged to set up enterprises on promising experimental ideas.
- Standardization of best practices from environmental, economics, technological and resource viewpoints shall we encouraged for application in the periphery of sustainable waste management.

Visualizing a framework for embedding indigenous knowledge in waste management practices

The practices involving the utilization of indigenous knowledge in sustainable waste management tend to be localized, concentrated and confined to a certain area. Streamlining such practices and making them diffused on a scale where their actual potential can be realized is a challenge before Indian policymakers. An institutional architecture is necessary to streamline and lay down indigenous knowledge domain in the realm of waste management practices. Therefore a multi layered structure of institutions entrusted with specialized function of utilizing indigenous knowledge in the waste management domain is essential to help effectively implement policies pertaining to utilization of indigenous knowledge in waste management.

National level architecture

At the national level a dual architecture of specialized executing agencies having expertise in the domain are necessary. The institutions working in rural, tribal and agricultural sector may push the use of traditional wisdom in sustainable waste management. Tribal Cooperative Marketing Development Federation of India Limited (TRIFED), Khadi and Village Industries Commission (KVIC), Agricultural and Processed Food Products Export Development Authority (APEDA), Agri-business clusters, mega food parks and such other specialized agencies such as the network of Indian Council of Agricultural Research(ICAR) may use indigenous knowledge in their operations of waste management to generate optimum value from the resources while minimizing the waste using indigenous knowledge.

Along with these executing agencies at the national level is required an innovation and network which can guide research, innovation and documentation to spur action at the national, regional and local level. In this regard institutions such as National Innovation Foundation and National Productivity Council may help to translate innovative ideas and traditional practices involving wisdom may help to combat the problem of growing waste (National Innovation Foundation n.d.). These institutions may adopt approaches of creative destruction to fuse modern scientific knowledge with traditional wisdom and evolve best practices which have state of the art standards.

Devising a strategic framework for infusing indigenous knowledge in waste management

A strategic framework for infusing indigenous knowledge in waste management operations is necessary to imbibe sustainability in waste management processes across different industries and sectors. The following steps may make use of indigenous knowledge in ensuring sustainability of waste management practices. Business intelligence can be used to optimize the strategic framework for infusing indigenous knowledge as it serves as a systematic frame of advancing projects (Srivastava 2018).

- Identification of sectors and processes where traditional knowledge is already being used to sustainably manage and recycle waste. For example various agricultural process and agro processing industries recycle waste to aggregate inputs for alternate supply chains of either energy or feed material industries. Demarcation of such sectors is imperative for scaling up use of indigenous knowledge to ensure long term sustainability of the waste management practices currently in use.

- Documentation of best practices where indigenous knowledge is being used to sustainably manage waste generating wealth from the West. These practices shall be converted into best practices and made available to industries for infusing them throughout.
- A strategic mission must be launched to identify new sectors such as construction, energy, textile etc. where indigenous knowledge may be tapped to infuse sustainability.
- Strategic mission to promote entrepreneurship in waste management domain imbibing indigenous knowledge shall be launched to promote such ventures developed by combining traditional wisdom and modern science to be provided with initial capital and hand holding.

Imbibing a culture for change

Waste management is not just about minimizing the environmental hazards or safe dumping of the waste remaining after a certain process. Rather it shall be seen as the residual resource to be used in other processes and have potential to develop a value chain and provide employment as well as entrepreneurship opportunities for the people involved in the value chain.

Therefore the term waste shall be replaced with, the term ‘residual’. This indication will signify that the remainder materials still hold value and are not wastes, which have alternative uses apart from the original process, required to get the desired output. As famous resource economist Erich Zimmerman opines “resources are not, they become”. Zimmerman tries to highlight that it is not the inherent properties of material that make them resources or waste rather it is our treatment that either makes a material a thing of great value or a thing of waste (Barnett n.d.).

Inspired from the traditional Indian knowledge of indestructibility of matter we must acknowledge that waste of one operation is a resource for another and hence a continuous exploration, deliberation, research and innovation shall drive the fusion of traditional wisdom and scientific knowledge to introduce changes in the value chain through value change. That is changes in our mindset to see discarded material as resources and residuals after a process and not waste will drive and enhancement of the Value chain management.

Creating platforms to integrate indigenous waste management in mainstream industries

For diffusing traditional wisdom and knowledge of waste management to different sectors and industries we need to create suitable structures where the industry leaders and people possessing indigenous knowledge of sustainable waste management culminate together to devise mechanisms to implement sustainable waste management practices in Indian industry. These actions will entail making indigenous stakeholders a part of the value chain and keeping a close connect with them will optimize the value obtained from industrial process and harmonize the residues with the natural environment of the surroundings. The traditional wisdom possessed by the local people can easily optimize waste management (Prakash 2005).

In order to create such structures, providing and enabling policy framework for commercial viability and hand holding of such ventures and platforms is necessary; herein lies the challenge for Indian policymakers to prepare such platforms and energize them with enthusiasm from the side of Indian industry. The industry must have a keen approach to integrate such practices with the ongoing waste management mechanisms and harness the potentialities offered by the indigenous knowledge.

Conclusion

Indigenous knowledge accumulates over long periods with passage of many civilizations assimilating knowledge and practice. In fact it is more of experiments and experiences emanating from repeated events that create such a pile. The last century has been a century of technological advancements that has pegged new forms of sciences on the central block. This has overshadowed indigenous treasure of solutions. Indigenous knowledge is covered with a blind sheet which prevents us to look at solutions as most of the experiences of indigenous knowledge have either not been documented but have existed in practice or if they have been documented a few languages like English are considered erudite thereby creating a blind towards such knowledge. Indigenous knowledge needs to be documented and then retried and retested which gives it credence in today's parlance and gets a common acceptance. Grass root innovation is an essential conduit to create successful solutions for sustainable living. We will have to put the commercial benefits arising out of new innovations on the back seat and connect with grass root indigenous knowledge solutions for a faster contribution to the globe.

Disclaimer

None

References

Barnett, Douglas E. (1888–1961) Zimmermann, Erich Walter. https://www.tshaonline.org/handbook/entries/zimmermann-erich-walter (accessed August 27, 2022).

Boojh, R. (2022). Waste Management and the Agenda 2030 in the Indian Context. In *Environmental Management in India: Waste to Wealth* (pp. 3-22). Springer, Cham.

Government of Haryana. (2018). "Galvanizing Organic Bio-Agro Resource." http://haryanadp.gov.in/Portals/0/GALVANIZING%20ORGANIC%20BIO-AGRO%20RESOURCE%20%28GOBAR-DHAN%29_1.pdf (accessed August 27, 2022).

Government of India. (May 12, 2021) *Waste to Wealth.* https://www.wastetowealth.gov.in/ (accessed August 26, 2022).

International Energy Agency. *Zero Defect Zero Effect (ZED) Certification Scheme.* https://www.iea.org/policies/2454-zero-defect-zero-effect-zed-certification-scheme (accessed August 22, 2022).

Liu, J., Shu, A., Song, W., Shi, W., Li, M., Zhang, W., ... & Gao, Z. (2021). Long-term organic fertilizer substitution increases rice yield by improving soil properties and regulating soil bacteria. *Geoderma*, *404*, 115287.

Kang, D. S. (2021). Environmental problems of the green revolution with a focus on Punjab, India. In *International dimensions of the environmental crisis* (pp. 191-215). Routledge.

Narayan, R. H. (2007). The conception of atom in greek and Indian Physics. *arXiv preprint arXiv:0712.0764*.

National Innovation Foundation. https://nif.org.in/dsd (accessed August 27, 2022).

Prakash, A., & Bahadur, A. (2005). From Revolutionary Peasants to Leaders of Export Revolution: A Saga of 1990's. *Vision*, *9*(2), 67-76.

Sarisky-Reed, Valerie. *Everyday Products Made from Biomass.* https://www.energy.gov/eere/articles/5-everyday-products-made-biomass-few-may-surprise-you (accessed August 24, 2022).

Srivastava, A., & Prakash, A. (2018). Leveraging Business Intelligence for Organizational Performance the Emerging Economy Context. *International Journal of Management Studies*, *5*(2), 1.

Surie, G. (2017). Creating the innovation ecosystem for renewable energy via social entrepreneurship: Insights from India. *Technological Forecasting and Social Change*, *121*, 184-195.

Tripathi, Anita: Prakash, Ajai: Prakash, Jyoti. "Economics and Market of Wastes." Chap. 14 in *Emerging Trends to Approaching Zero Waste*, by Lalit Goswami: Chaudhry Muntashir Hassan: Singh Sunpreet:, 319-338. Lucknow: Elsevier, 2022.

Chapter 2

Role of microbes in remediation of waste and sustainable green environment

Avishka Singh, Faiza Fatima, Rachna Chaturvedi*
Amity Institute of Biotechnology,
Amity University Uttar Pradesh, Lucknow

Abstract

Natural environment security and manageability are one of the main pressing issues in the present situation. The constant arrival of hazardous waste and foreign substances because of ill-advised industrialization and uncontrolled urbanization has the most prominent danger to humankind in the short as well as long term and creates a huge burden on real assets. The reduction of these toxins is of the most extreme significance for a sustainable green environment. Therefore it is very important to do improvement and remediate the contaminated environment in such a manner that makes a viable and comfortable environment for our upcoming generations. The impacts of organisms on their climate can be gainful or destructive or apparent concerning human measure or perception. The distinctive character of microbes can be utilized successfully for revitalizing the natural environment. The main impact of the microorganisms on earth is their capacity to reuse the essential components that make up every single living framework, particularly carbon, oxygen, and nitrogen (N). Essential production includes photosynthetic living beings which take up carbon dioxide from the climate also, convert it to natural (cell) material. Right now, various techniques and systems are applied in the region in various parts of the planet. Cleaning of polluted climate in a manageable manner is a lot of vital; in such manner, the job of microorganisms in waste management and biodegradation of pollutants has heightened. Unique microorganism-based biotechnological devices like bioremediation, biodegradation, composting and biotransformation have been utilized together and degrade the extent of toxins. So, the utilization of microorganisms with various technologies is present to treat various squander, as well as being an eco-accommodating, savvy, and earth-feasible

* Corresponding Author's Email: rchaturvedi1@lko.amity.edu

strategy. The important target of this chapter is to propose furthermore, advancement in the utilization of a generally reasonable and climate amicable strategy for treating polluted waste with the utilization of various microbial agents towards ecological maintainability and sustainability in a greener way

Keywords: Microbes, Remediation, Waste, sustainable development, green environment

Introduction

The global environment is stressed due to speedy development, business modernization, and the enormous strain of inhabitants on the environmental reserves. Several irrational behaviors of contemporary culture produce enormous waste which takes the lead in pollution (Raj et al. 2018a, b). Furthermore, the enhanced amount of utilization of unprocessed substances due to industries' products in the unloading of enormous amounts of chemical and radioactive wastes in the atmosphere, advances to irreversible destruction of the global atmosphere. The production of waste is the most important cause of a deficiency in resources and power and increases ecological concerns and expense on the public for its compilation, management, dumping, and total administration (Jhariya et al. 2018). Production of waste in our country is immediately associated with and directly proportional to the development of urbanization and industrialization enhancement (Sharma and Shah 2005). The situation of waste management is considerably distinction between established and emerging countries as growing countries are lacking good strategies for waste management. So, waste management has to turn out to be an exceptionally important aspect and acknowledges significance related to environmental deprivation and sustainability(Brewer 2001). Uncontrolled, unsystematic, and natural properties of elimination of waste products on the borders of townships and boundary of rural communities mark the abundance of landfills, which are almost irreversible means to return to an appropriate condition but likewise responsible for ecological consequences concerning soil and groundwater contamination and involvement of global warming. An efficient or operational approach to Waste Management is the necessity of the present scenario and must be cost-effectively and ecologically viable. Several common techniques of Waste management, applied consistently are Incineration, Sanitary landfills, Recycling and Avoidance and reduction.

Maintenance of contaminated atmosphere in a natural way is extremely required; in this respect, the role of microbes in Waste management and biodegradation of pollutants has increased in current days (Banerjee et al. 2018). Various microbe created biotechnological implements such as biotransformation, biodegradation bioremediation, and bio composting have been utilized to gather and decompose an enormous variety of pollutants successfully (Maghraby and Hassan (2018) stated that Cladophora sp. (green algae) has a high-level bioaccumulation capacity for noxious metallic elements and can be utilized as a powerful substitute for Waste management representatives. Furthermore, microbial ecology is the most important component for the appropriate implementation of natural methods of waste management. (Martinez et al. 2018) exhibit that archaea and bacteria are leading in the bioreactors of the Successful Role of Microbes in Waste Management and Ecological Waste management in the polar arctic region in Finland. Currently, nanoparticles have been employed successfully to improve the action of microorganisms, characterized as nano bioremediation. In the United States of America, Deinococcus radiodurans (extremophilic bacteria) is utilized in radioactive waste elimination approaches as it is a radioactive-tolerant organism and can resist radiation easily (Varma et al. 2017). Therefore, the application of microbes with various biotechnologies is an extremely efficient technique to manage different wastes, in addition to an eco-friendly, cost-effective and ecologically viable approach.

Microbes are broadly disseminated on the planet of their metabolic capability is extremely notable and they can simply flourish in a broad variety of ecological circumstances.

Waste and its Classification

Waste is undesirable or unusable materials, that should be disposed of with proper care, may also be defined as the trash found in our environmental elements. Trash is predominantly viewed as solid waste that is produced from our homes (domestic waste), squanders from schools, workplaces, metropolitan endlessly waste from ventures, and production lines (industrial waste). Waste is produced from human action for the most part. Generally, the biosphere degenerated quickly because of the consistent influx of perilous toxins from various industries throughout the world. Main categories of waste are as soli, dry, wet, degradable and nondegradable (Fig 1)

Figure 1. *Categories of waste*

Waste can be of any physical state liquid, solid, or gas. Waste is characterized by its source and by its qualities as industrial waste, agricultural waste, commercial waste, and domestic waste

Some normal attributes utilized in the grouping of waste incorporate the actual states, actual properties, reusable possibilities, biodegradable possibilities, the wellspring of production, and the level of environmental effect may be as Hazardous and non-hazardous waste

Because of the restricted extent of the assessment study, fluid squanders, which can be discarded through sewer organizations or lost to groundwater, and risky squander, which require more tough natural controls, in light of their potential to impose damage, are prohibited.

Waste Management

Human associations with the environment (human exercises) have consistently brought about waste production. Waste management is essentially the capacity, assortment, and removal and supervising the waste materials.

Waste management is an interaction by which waste is accumulated, moved, and handled before the removal of any excess deposits. The fundamental point of waste administration is to diminish the impacts and results of waste on human well-being and climate. It keeps on being a challenge with quick urbanization, modern insurgency, and gigantic tension of the populace that put incredible stress on the worldwide climate.

Waste management has been categorized into four parts –electronic, municipal, industrial, and biomedical- and these wastes are administered by specific arrangements. The waste management principle is based on the 4R theory – refuse, reduce, reuse and recycle. A portion of the normally utilized waste management strategies are landfills, burning, treating the soil, and gasification.

Role of Microbes in Waste Management

Like any assortment of microorganisms, organisms found in wastewater can be set up in significant gatherings based on specific qualities. Microorganisms share a few essential elements of their sort. The significant gatherings of microorganisms are protozoa, parasites, green growth, and microscopic organisms. Microbial biotechnology in the management of waste the during the use of present-day logical devices and procedures which utilize a wide assortment of microorganisms in controlled conditions without upsetting the biological system. Numerous microorganisms are extremely effective to oxidize or decompose various organic compounds (OC) into a further easy and steady end item for consumption

(Atalia et al. 2015). Most normal and productive strategies embraced in different degrees of waste management are treating the soil as Bio composting, biodegradation, bioremediation and biotransformation.(Fig 2).

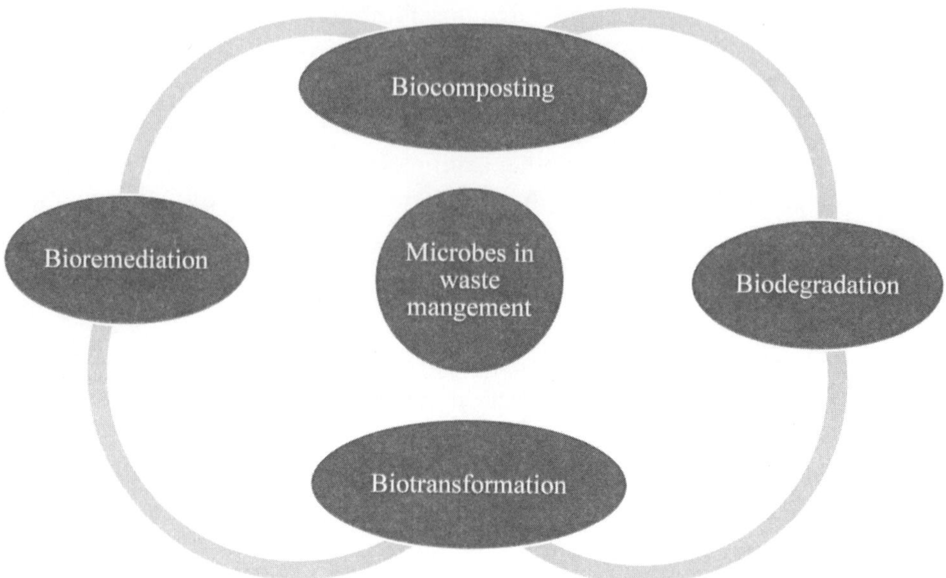

Figure 2. *Microbe and different processes involved in waste management*

Subsequently, the most satisfactory solution for the created waste generated in the climate is with the end goal that will advantageously incorporate them back into the climate. A few instances of the utilization of microorganisms in the climate are examined as follows:

Soil treatment: The nitrogen cycle in soils relies upon the fixation of nitrogen present in the atmosphere. This is accomplished by various diazotrophs. One way this can happen is in the knobs in the foundations of vegetables that contain cooperative microscopic organisms of the genera *Rhizobium, Mesorhizobium, Sinorhizobium, Bradyrhizobium*, and *Azorhizobium.*

Sewage treatment: Most of all oxidative sewage treatment processes depend on a huge scope of microorganisms to oxidize natural constituents which are not manageable to sedimentation or buoyancy. Anaerobic microorganisms are additionally used to diminish slime solids delivering methane gas (among different gases) and a sterile mineralized buildup. In consumable water treatment, one strategy, the sluggish sand channel, utilizes a complex coagulated layer made out of a wide scope of microorganisms to eliminate both disintegrated and particulate material from crude water.

Energy generation: Microorganisms are utilized in aging to create ethanol and in biogas reactors to deliver methane. Scientists are exploring the utilization of green growth to create fluid fuels, and microbes to change over different types of rural and metropolitan waste into usable fuels.

Bio Composting

Composting is a high-impact disintegration process and is worked by a different populace of microorganisms. This interaction has been generally rehearsed for various kinds of waste through the metabolic action of the microbial consortium. Treating the soil has been utilized to change and settle natural waste into a more secure and balanced structure that can be utilized in different agrarian practices. It is a monetarily and ecologically fitting strategy for taking care of waste. Bio composting supports to improve full nutrient management and the land usage of compost might be further to reduce soil organic matter decrease and soil erosion (Van Camp et al. 2004) The main product of composting is humus produced by the interaction of water, air. Organic matter micro and macro organisms and the by products are Compost, heat, CO_2 and water (Fig 3) Fertilizing the soil happens in three unique stages: mesophilic stage, thermophilic stage, and cooling and development stage. Two elements direct the spanner on the other hand length of the fertilizing the soil stages: the kinds of fertilizing the soil natural matter (OM) and the productivity or adequacy of the cycle determined by the level of air circulation and unsettling.

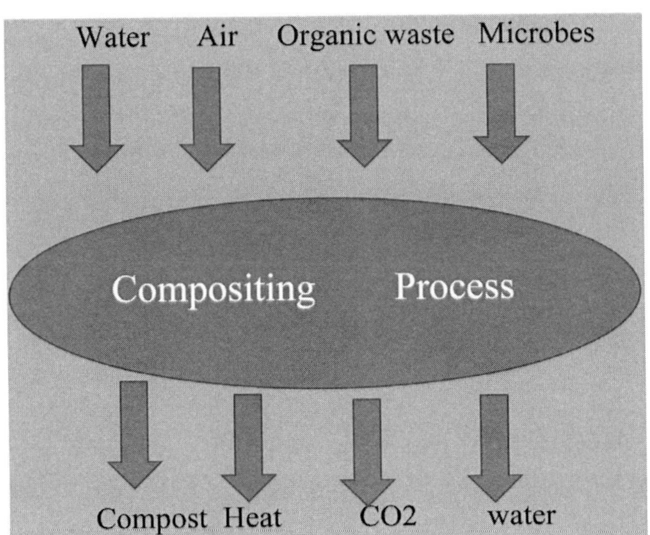

Figure 3. *Process of bio composting In puts and by products*

Bio Composting Factors

Microbes

A few microorganisms are especially powerful to oxidize or decay different natural compounds into more straightforward and balanced results. Microorganisms can fill in the temperature range somewhere in the range of 10 and 45 °C and debase degradable parts. The thermophilic stage is the dynamic period of treating the soil and can keep going for quite a long time. In the thermophilic stage, a large portion of the Organic matter is corrupted. These may be categorized in the following types

Third level Decomposers Bigger creatures, as Macro-organisms, substantially break down the organic material into minor pieces by grinding, splitting and, slurping to produce compost Examples of this category are centipedes, woodlice (sow bugs) slugs, snails, spiders, flies, beetles, ants, and "composting" worms,

Second level Decomposers Nematodes, Dspringtails, mold mites and protozoa, eat the organic matter and made them available for the the organisms of first level decomposers.

First level decomposers Smaller Microbes, play the major part in the process of composting. Microbes utilize the organic matter in the compost waste bin or pile as a supply of food following in its decomposition to the rich brown compost.

Moisture content

To begin manure the appropriate and ideal moisture content is for the most part 60-70%. Be that as it may, at the last stages, 50-60% is the ideal dampness content. Dampness content higher than 75% and lower than 30% fundamentally diminishes microbial exercises. A legitimate harmony between microbial movement and accessible oxygen successfully deals with moisture content. Overabundance dampness makes anaerobic circumstances which bring about unfortunate items and awful smells.

Particle Nature and Size

Nature and the size of particles are vital for fertilizing the soil. Molecule size influences oxygen accessibility into the heap as well as microbial entrance to the substrate. Particles that are more modest in size increment the surface region required for microbial assault, and particles that are bigger limit the surface region accessible for microbial assault which prompts delayed down or on the other hand even stops the treating of the soil hardware.

Temperature

It has been accounted for those good temperatures for fertilizing the soil to happen at the scope of 52-60 °C. During treating the soil, the greatest temperature goes up to 60-70 °C, where the majority of the microorganisms are not as dynamic. Treating the soil cycle can be stopped or altogether consistent at a temperature that is under 20 °C. It has additionally been accounted for that a temperature over 60 °C can decrease microbial action since the temperature at that level crosses the ideal thermophilic range of microorganisms.

Biodegradation

Biodegradation is a natural method of corruption of synthetic compounds (Focht DD,2020) In this interaction living microbial creatures are utilized to debase natural substances into more modest mixtures. Biodegradation is firmly connected with WM and ecological remediation. As far as microbial science, biodegradation is the corruption of organic components by an enormous variety of microbial populace predominantly microscopic organisms, yeast, and parasites. Various sorts of microorganisms engaged in the biodegradation process.

Microbial Population involved in Biodegradation

There are many microorganisms (especially bacterial and fungal foundations) that have a mechanism to degrade huge and complex hydrocarbons into simpler biomolecules. They are specifically Gram-positive and Gram-negative in addition to some species of fungal foundation like *Aspergillus* species. Other species of microbes like Streptococcus, *Micrococcus* (Gram-bad), *Staphylococcus, Moraxella*, and *pseudomonas* (Gram-high quality) and species of fungi (*Aspergillus glaucus* and *Aspergillusniger*) are also involved in the biodegradation device. Microorganisms produce positive extracellular enzymes and thereby break down complicated polymers into simpler molecules to further use them as carbon and energy assets. Organisms can use degraded plastic as a source of carbon.

Plastic Waste Biodegradation

With increasing plastics being employed in human lives and increasing stress being placed on capacities to be had for plastic waste disposal, the want for biodegradable plastics and biodegradation of plastic wastes has assumed growing importance within the previous couple of years. This evaluation looks at the technological development made in the development of extra without difficulty biodegradable plastics and the biodegradation of conventional plastics utilizing microorganisms. Additives, consisting of seasoned oxidants and starch, are implemented in synthetic substances to alter and make plastics biodegradable. Recent research has shown that thermoplastics derived from polyolefins, historically taken into consideration as resistant to biodegradation in ambient surroundings, are biodegraded following photograph degradation and chemical degradation. Thermoset plastics, consisting of aliphatic polyester and polyester polyurethane, are without problem attacked by way of microorganisms immediately because of the capacity of hydrolytic cleavage of ester or urethane bonds in their structures. Some microorganisms were isolated to make use of polyurethane as a sole source of carbon and nitrogen source. Aliphatic-fragrant polyesters have active business applications because of their appropriate mechanical properties and biodegradability.

Plastics are nearly nondegradable and hugely dumped after getting used. Because of this, the widespread application of plastic can reason an extreme environmental chance to one-of-a-kind ecosystems. Terrestrial and marine habitats are substantially hampered due to the accumulation of plastic materials. Disposal of plastic is a primary situation due to its high production and extensive usage. Plastic degradation releases a completely fatal greenhouse gas, methane, which brings global warming.

Marine surroundings are also affected by employing plastic pollution as it could cause morbidity to many marine animals. Different research implies that many marine animals which include sea turtles, cetaceans, zooplankton, and sea chickens frequently ingest plastics that can be continuously dumped within the aquatic system. When polyvinyl chloride (PVC) is burned, it releases dioxins which are a powerful carcinogen and also an immune and reproductive system disrupter. Some plastic-made bottles have bisphenol as a constituent that is verified as a carcinogen and also hampers the physiological system of humans.

Seven Types of Plastics

- High-Density Polyethylene (HDPE)
- Low-Density Polyethylene (LDPE)
- Polyethylene Terephthalate (PETE or PET)
- Polystyrene or Styrofoam (PS)
- Polyvinyl Chloride (PVC)
- Polypropylene (PP)
- Miscellaneous plastics (includes: polycarbonate, polylactide, acrylic, acrylonitrile butadiene, styrene, fiberglass, and nylon)

Microbes have considerable responsibility in degradation and festering of artificial and pure plastics (Meena et al. 2017). Microbial biodegradation of plastic is sustainable approach for tackling environmental threats facing a big issue due to use of plastic everywhere. (Fig 4).

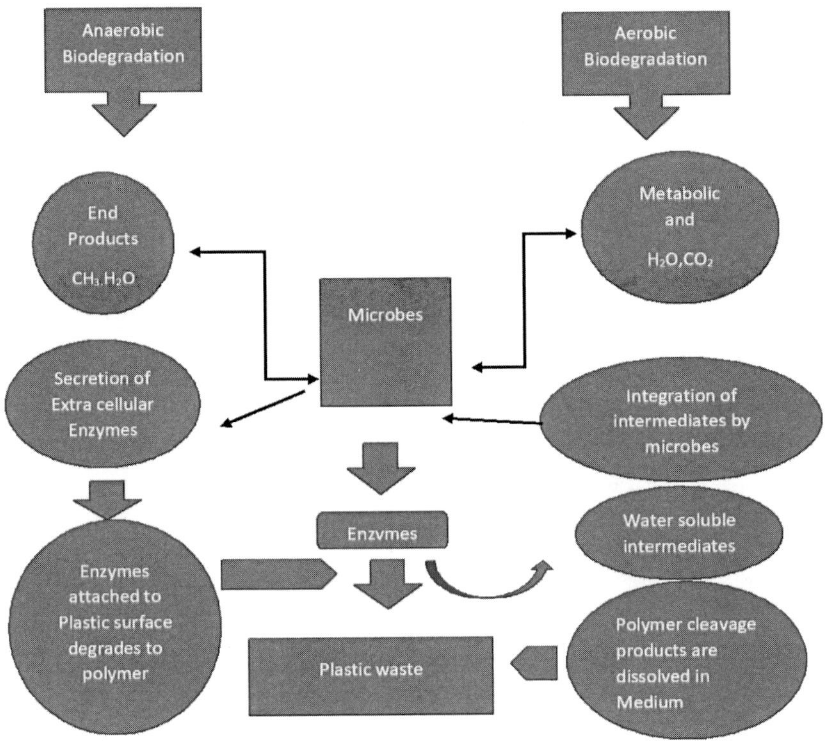

Figure 4. *Microbial degradation scheme for plastic degradation.*

Xenobiotic Compounds Biodegradation

Xenobiotic compounds are human-made chemical compounds that might be overseas to the character. They are extraordinarily thermodynamically stable, consequently are particularly persisting in the environment. The main sources by which xenobiotics are launched into the surroundings are chemical and pharmaceutical industries which generate kinds of xenobiotic and synthetic polymers. Modernization of agriculture and mining produces massive quantities of fertilizers and pesticides and releases heavy metals into biogeochemical cycles. Toxic chlorinated compounds launched from the pulp and paper enterprise and accidental oil spillage in large amounts make the state of affairs more catastrophic. Xenobiotic substances are exceedingly new to the environment, subsequently now not easily removed. Xenobiotics are poisonous and show an unfavorable effect on public health inclusive of disrupting numerous mobile pathways which might be accountable to alter development and increase. Xenobiotic can put an unfavorable effect on lower and higher eukaryotic organisms.

Xenobiotic is tough to break down and degrade due to its recalcitrant nature. For breaking down such compounds, distinctive microorganisms and their enzymes play a big role.(Fig 5) Numerous studies suggest that in the assessment of yeast and fungi, bacteria are the primary microorganisms for xenobiotic detoxing. The role of protozoa and microalgae could be very low in xenobiotic degradation. *Pseudomonas* sp. And *bacillus* sp. Are amazing and effective in the degradation of xenobiotics the various bacterial organization. Earlier studies have also been reported degradation of various xenobiotic compounds and heavy metals by the members of genus *Bacillus* (Birolli et al., 2016; Upadhyay et al., 2017; Arora et al., 2018; Díez-Méndez et al., 2019).

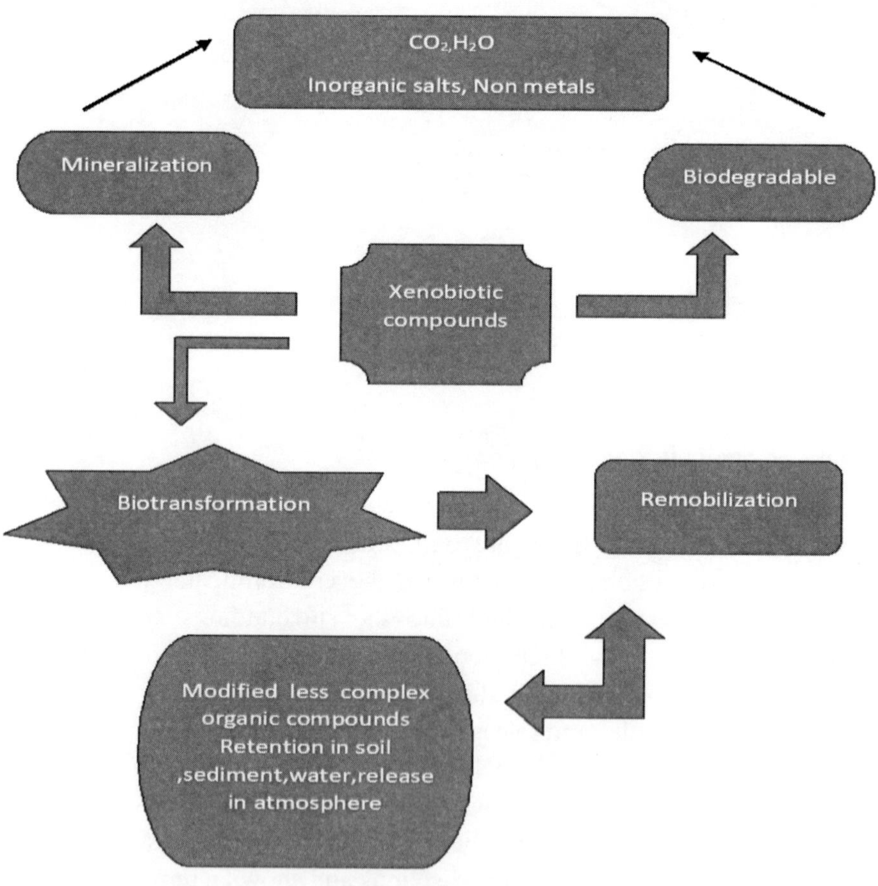

Figure 5. *Microbial Degradation of xenobiotic compound*

Bioremediation

Bioremediation is a normal practice which involved in the use of microbes to eliminate waste or contaminant. This is an atmosphere-responsive and sustainable approach as it includes eco-responsive microorganisms in handling the waste (Kensa 2011). I. Remove debris or waste material from water and soil. it can be eco-friendly and proprietary techniques as they involve environmental microbes in solids processing waste. Various microbes including aerobic, anaerobic, and fungal bacteria are involved in the bioremediation process. Bioremediation is heavily involved in the degradation, disposal, immobilization, or detoxification of various chemical wastes and physically hazardous materials from the environment through the holistic action of microorganisms. The basic principle is to break down pollutants and convert them into less toxic forms. There are two types of factors - biotic and abiotic conditions - that determine the rate of degradation. Various methods and strategies are currently used for bioremediation processes. Bioremediation is highly involved in degradation, eradication, immobilization, or detoxification diverse chemical wastes and physical hazardous materials from the surrounding through the all-inclusive and action of microorganisms (Fig 6)

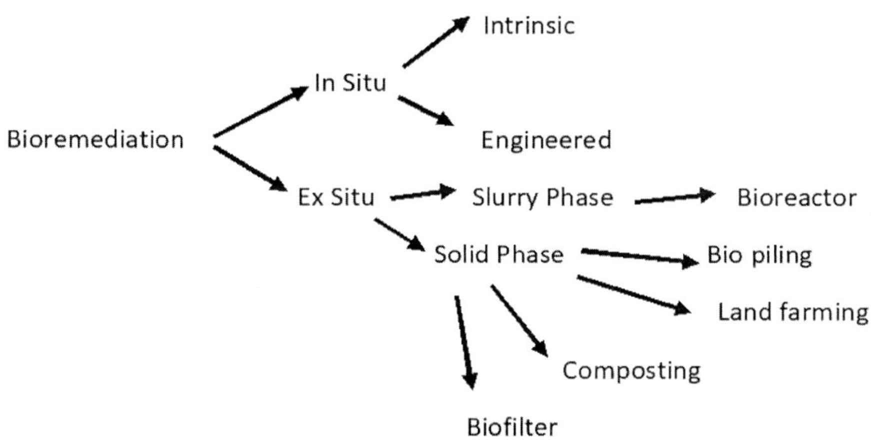

Figure 6. *Bioremediation approaches for sustainable environment*

The mechanism of bioremediation is to reduce, detoxify, degrade, mineralize, or convert more toxic pollutants into less toxic ones. The pollutant removal process mainly depends on the type of pollutant, which includes pesticides, agrochemicals, chlorinated compounds, heavy metals, xenobiotic compounds, organic halogens, greenhouse gases, hydrocarbons, nuclear waste, dyes, plastics, and sludge. Purification techniques are applied to remove toxic waste from the polluted environment. Bioremediation is heavily involved in the degradation, disposal, immobilization, or detoxification of various chemical wastes and physically hazardous materials from the environment through the holistic action of microorganisms.

Microbial Populations Involved in Bioremediation

The bioremediation control and optimization process is a complex system consisting of many factors. these factors include the presence of a microbial population capable of doing pollutant degradation, pollutant availability increases microbial population, and environmental factors (soil, temperature, pH, presence of oxygen or other electron acceptors and nutrients). Microorganisms can be isolated from almost anyone's environmental conditions. Microbes adapt and grow in Sub-zero temperatures and extreme heat, in desert conditions in water with excess oxygen and anaerobic conditions, and the presence of dangerous a source of energy and a source of carbon. The adaptability of microbes and other biological systems can be used to reduce or eliminate environmental hazards.

We can divide these microorganisms into groups:

Aerobics: In the presence of oxygen. Examples of aerobic exercise are bacteria known for their ability to degrade *Pseudomonas, Alcaligenes, Sphingomonas*, *Rhodococcus*, and *mycobacteria*.[6] This microbe is frequently reported to break down pesticides and hydrocarbons, both alkanes and polyaromatic Compounds.

Anaerobic: In the absence of oxygen. are anaerobic bacteria are not used as often as aerobic bacteria. There is increasing interest in anaerobic bacteria used forBiphenyl polychlorinated (PCB) bioremediation in riversSediment, trichloroethylene solvent dichlorination (TCE) and chloroform.

Ligninolytic Fungus: Fungus like white rot fungus *Phanaerochaete chrysosporium* has the ability to a very wide range of resistant or toxic environmental pollutants. The substrates commonly used are straw, sawdust,

Methylotrophs: Aerobic bacteria that grow on methane for carbon and energy. The early enzyme in the pathway for aerobic degradation, methane monooxygenase, has a broad range of substrates and is effective against a broad spectrum of compounds including chlorinated aliphatic trichloroethylene and 1,2-dichloroethane.

Biotransformation

Biotransformation is the conversion of toxic compounds into less ductile compounds in less toxic form. Bacteria and fungi are the main groups of microorganisms and your enzymes are involved in this process.

Microorganism cells are essential for biotransformation for several reasons, such as:

- Surface to volume ratio: High surface-to-volume ratio in the case of microorganism-mediated biotransformation.
- Microbial Cell Growth Rate: Microbial cells have a high growth rate, which minimizes the duration of biomass conversion.
- Metabolic Rate: The metabolic rate in a microorganism is very high and necessary for an effective transformation.
- Sterile conditions: For effective biotransformation, it is necessary. Maintains sterile conditions of microorganisms.

Microbial Electro-Remediation

Due to human activities and rapid industrialization, a large amount of metallic wastewater is generated. Water contaminated with metals (loids) poses major health and environmental hazard due to their toxicity potential and carcinogenic effects, strict guidelines have been put in place for the disposal of various. Metal ions in wastewater to avoid environmental pollution. For example, based on the US The EPA lists a priority pollutant, naturally occurring arsenic as a highly toxic metalloid and disaster, more than 250 million people face arsenic poisoning worldwide, mainly in developing countries, resulting in neurological impairment and even death. In perfect treatment, the approach is not to remove metal but to build and restore it during the treatment process. Microbial electro-remediation tools those are targeted in relation to the development of the metabolic action of electrochemically involved biocatalyst by providing organic/inorganic nutrients, electron acceptors, or donors, therefore encouraging oxidation or reduction of pollutants. In this respect, BEC has achieved much attention, in which the management of environment is achievable by methods of electrical current/ voltage that works as a donor or acceptor for hazardous waste remediation. Indeed, this is necessary to provide solutions to traditional techniques such as Coagulation, precipitation, and adsorption involved in metal removal, which are generally considered deficient effective. Adsorption, ion exchange, and membrane technology approach based on activated carbon are very expensive to treat because it requires a large amount of waste and industrial waste with small amounts of heavy metal ions, this technique is not applicable on a large-scale industry scale.

Waste Water Treatment

Access to clean water is considered a fundamental human right that is denied by most of the population of the world's population, and on top of that, the availability of clean water has a huge impact on people's Development. Urban population is expected to double to over 5 billion by 2030 around the world and such problems related to water pollution if left unchecked, will grow with unavoidable public health side effects. It is therefore important that this is where the economy is at its worst as affected countries grow, part of their wealth will be spent on sanitation infrastructure. The issue of major concern is the quantity and quality of waste generated and discharged into natural waters. Agencies have dominated the headlines of various international media and the richest countries are investing. Significant efforts have been made to limit the pollution of large populations. world report heavy metal contamination of groundwater has become a serious public health problem this context, adequate knowledge of wastewater sources is essential is important to identify its biological, chemical, and physical aspects and carry out an appropriate treatment strategy. For this reason, various methods of water purification technology are constantly being developed and improved over time as research facilities around the world improved. Associated with relevant waste source classification, water treatment technology can also be divided into biological, chemical, and physical treatment techniques and are further classified as in situ or ex situ technology. To better explain the application of technology in situ and ex-situ, in situ. Corrective action means maintenance performed at a specific location while ex situ maintenance is included the Removal of pollutants in remote locations. Flocculation is an excellent example of chemical water. Handling measures used to treat surface water and groundwater on site.

In wastewater treatment plants, wider substrate biodegradation is relatively better than pure culture, because the microorganisms are arranged in a species-rich structure. In the biological wastewater treatment process, most of the microorganisms are present in the microbial aggregate Forms, including biofilms, flakes, and granules. The application of different electron microscopic methods shows the presence of complex mixtures of high molecular weight polymers, called extracellular polymer substances in pure culture, activated sludge, biofilm, and granular sludge. Microbial EPS Surfaces play an important role in holding microbial aggregates together in a three-dimensional gel-like hydrated matrix through physio-chemical interactions which are weak. This weak power of the interactions includes hydrogen bonding, electrostatic, hydrophobic, and van der Waals forces. (Wang L. et al. 2020). EPS is a Metabolic by product that accumulates on the surface of bacterial cells (Angelin J, and Kavitha M, 2020)

They are formed from microorganisms present in the bioreactor that consume the organic matter present in the wastewater. Various mechanisms such as secretion, excretion, cell lysis, and absorption of waste components. Water contributes to the accumulation of EPS From a technical point of view, EPS has several basic properties such as biosorption potential and biodegradability. EPS consists of various organic macromolecules including polysaccharides, proteins, nucleic acids, phospholipases, and other low molecular weight non-polymer components. Also, organic substances such as humic substances. It has also been reported that uronic acid is present in EPS.

Microorganism and Waste water treatment

Bacteria are of utmost significance in the treatment of Waste water. Most of them are facultatively residing in both aerobic or anaerobic environments (Absar 2005). Alcaligenes sp., Pseudomonas sp., Flavobacterium sp., Citramonas sp., Zoogloe sp. Arthrobacter sp., Achromobacter sp., and Acinetobacter sp. are a few significant bacteria observed in Waste water treatment scheme (Oehmen et al. 2007) Filaments of filamentous bacteria support the particular mass to develop and avoid the disintegrate activity in the treatment procedure. Filamentous bacteria can trigger resolution difficulties if they are discovered in extreme numbers. (Paillard et al. 2005). Algae is a powerful creature to cleanse Waste water naturally as these are capable to collect heavy metallic elements, organic and inorganic pollutants and also pesticides. Microalgae have achieved a plenty of significance in the treatment of Waste water (Lloyd and Frederick 2000). Some of the fungal species are capable of oxidizing ammonia to nitrite. Zoogloea sp. And Sphaerotilus natans are the best for sewage treatment (LeChevallier and Au 2004). Different types of filamentous fungi are observed indeed in Waste water treatment methods. Fungi, like Fusarium sp. Absidia sp., Penicillium sp., and Aspergillus sp., are good to eliminate Carbon and nutrient resources in Waste water (Akpor et al. 2013). Furthermore, a few of fungi take advantage of their fungal hyphae for catching and adsorbing floating artifacts to alleviate their strength and nutritional requirement. It has been described that a variety of filamentous fungi produce a few enzymes which support to degrade the substrates throughout Waste water treatment methods (Molla et al. 2004; Buragohain et al. 2017).

Conclusion

Environmental protection and sustainability are recognized as top priorities that deserve critical attention worldwide as they are crucial to future progress. To ensure sustainability, waste management, conservation and biodiversity, and handling of pollutants and contaminants are key areas that are in dire need of priority. Nowadays, protecting the environment from decay is not only about removing pollutants and pollutants, but also about recycling and reusing hazardous materials by converting various wastes for prosperity. Aesthetically and ecologically useful. Interested in using various microorganisms has been increased and prioritized in the last scenario while humanity struggles to find sustainable ways to eliminate polluted environments and waste. With the advent of biotechnology, the potential of microorganisms for specific uses has received increasing attention and speculation. Microorganisms are unique and often unpredictable. Microorganisms can be used as an effective way to solve many environmental problems. The scientific and reliable use of microbes is unhealthy and reveals a spectacular development of scientific research and innovative tools to offer an efficient way. Protection of our planet and modern means of biological mass destruction and environmental monitoring. It can be concluded that the use of microorganisms and microbiology with Engineering opens up new perspectives in the field of sustainable development, especially in the field of the environment and other important environmental issues.

Meanwhile, Waste management is getting more and more important all over the world, and also in India is an integral part of environmental and sustainable development. India has a great opportunity, and it is very true to say that the success is very small. So certain areas need to be answered to understand a meaningful future Waste Management. Rapid urbanization and industrialization put great pressure on Wastewater around the world. So the principle of Waste Management should be based on prudence and sustainable development. Keys to effective Waste Management must be identified to ensure the separation of waste from its source and also must go through different. Types of recycling and recovery before landfill. Biotechnology and microorganisms can be used effectively in Waste Management, hence more research and knowledge of the role and application of biotechnology and role of microorganisms will be developed, and Energy production from waste and waste for compost will minimize waste. Biodegradation is believed to be part of the waste somehow separated; the challenge is effectively reduced. Future goals should be to evolve and redesign the long-term vision and strategy to deal with its changing lifestyle. Community participation and knowledge development and Resilience have a direct impact.

Disclaimer

None

References

Absar AK, Water and wastewater properties and characteristics. In: Water encyclopedia: domestic, municipal and industrial water supply and waste disposal. (2005) Wiley, New Jersey, pp 903–905

Akpor OB, Adelani-Akande T, Aderiye BI The effect of temperature on nutrient removal from wastewater by selected fungal species. Int J Curr Microbiol Appl Sci (2013) 2(9):328–340

Angelin J, and Kavitha M, Exopolysaccharides from probiotic bacteria and their health potential Exopolysaccharides from probiotic bacteria and their health potential" International journal of biological macromolecules 162, (2020): 853-865. doi: 10.1016/j.ijbiomac.2020.06.190

Arora, P. K., Srivastava, A., Garg, S. K., and Singh, V. P. Recent advances in degradation of chloronitrophenols. Bioresour. Techol. (2018). 250, 902–909. doi: 10.1016/j.biortech.2017.12.007

Atalia KR, Buha DM, Bhavsar KA, Shah NK Review on composting of municipal solid waste. IOSR J Environ Sci Toxicol Food Tech (2015) 9(5):20–29

Banerjee A, Jhariya MK, Yadav DK, Raj A Micro-remediation of metals: a new frontier in bioremediation. In: Hussain C (ed) Handbook of environmental materials management. (2018) Springer. https://doi.org/10.1007/978-3-319-58538-3_10-1, ISBN:978-3-319-58538-3.

Brewer LJ Maturity and stability evaluation of composted yard debris. M.Sc. thesis, Oregon State University, Corvallis, USA(2001)

Birolli, W. G., Borges, E. M., Nitschke, M., Romão, L. P., and Porto, A. L. Biodegradation pathway of the pyrethroid pesticide esfenvalerate by bacteria from different biomes. Water Air Soil Poll. (2016). 227:271. doi: 10.1007/s11270-016-2968-y CrossRef Full Text | Google Scholar

Buragohain S, Sharma B, Nath JD, Gogaoi N, Meena RS, Lal R Impact of ten years of bio-fertilizer use on soil quality and rice yield on an inceptisol in Assam, India. Soil Res. (2017) https:// doi.org/10.1071/SR17001

Díez-Méndez, A., García-Fraile, P., Solano, F., and Rivas, R. The ant lasius niger is a new source of bacterial enzymes with biotechnological potential for bleaching dye. Sci. Rep. (2019). 9:15217. doi: 10.1038/s41598-019-51669-wPubMed Abstract | CrossRef Full Text | Google Scholar

Focht DD. "Biodegradation". AccessScience. doi:10.1036/1097-8542.422025(2020).

Jhariya MK, Yadav DK, Banerjee A (2018) Plant mediated transformation and habitat restoration: phytoremediation an eco-friendly approach. In: Gautam A, Pathak C (eds) Metallic contamination and its Toxicity. Daya Publishing House, A Division of Astral International Pvt. Ltd, New Delhi, pp 231–247, ISBN:9789351248880)

Kensa VM Bioremediation-an overview. J Ind Pollut Control (2011) 27(2):161–168

Lloyd BJ, Frederick GL Parasite removal by waste stabilization pond systems and the relationship between concentrations in sewage and prevalence in the community. Water Sci Technol (2000) 42(10):375–386

LeChevallier MW, Au K Inactivation (Disinfection) processes. Water and treatment and pathogen control. IWA Publishing, (2004) pp. 41–65

Maghraby DM, Hassan J, Heavy metals Bioaccumulation by the green alga Cladophora herpestica in Lake Mariut, Alexandria, Egypt. J Pollut (2018) 1:1

Martinez AG, Sihvonen M, Palazon BM, Sanchez AR, Mikola A, Vahala R Microbial ecology of full-scale wastewater treatment systems in the polar Arctic circle: Archaea, Bacteria and Fungi. (2018) Sci Rep 8:2208. https://doi.org/10.1038/s41598-018-20633-5

Meena RS, Kumar S, Pandey A Response of sulfur and lime levels on productivity, nutrient content and uptake of sesame under guava (Psidium guajava L.) based agri-horti system in an acidic soil of eastern Uttar Pradesh, India. (2017) J Crop Weed 13(2):222–227

Molla AH, Fakhrul-Razi A, Alam MZ Evaluation of solid-state bioconversion of domestic wastewater sludge as a promising environmental friendly disposal technique. Water Res (2004) 38(19):4143–4152

Oehmen A, Lemos C, Carvalho G, Yuan Z, Keler J, Blackall LL, Reis AM Advances in enhanced biological phosphorus: from micro to macro scale. Water Res (2007) 41:2271–2300 PubMed Abstract | CrossRef Full Text | Google Scholar

Paillard D, Dubois V, Thiebaut R, Nathier F, Hogland E, Caumette P, Quentine C Occurrence of Listeria spp. In effluents of French urban wastewater treatment plants. Appl Environ Microbiol (2005) 71(11):7562–7566

Raj A, Jhariya MK, Bargali SS Climate smart agriculture and carbon sequestration. In: Pandey CB, Gaur MK, Goyal RK (eds) Climate change and agroforestry: adaptation mitigation and livelihood security. New India Publishing Agency (NIPA), New Delhi, (2018a) pp 1–19, ISBN:9789-386546067

Raj A, Jhariya MK, Harne SS Threats to Biodiversity and Conservation strategies, pp 304–320. In: Sood KK, Mahajan V (eds) Forests, climate change and biodiversity, Kalyani Publisher, (2018b) New Delhi, India, 381 p

Sharma S, Shah KW Generation and disposal of solid waste in Hoshangabad. In: Proceedings of the second International Congress of chemistry and environment, Indore, India, (2005) pp. 749–751

Upadhyay, N., Vishwakarma, K., Singh, J., Mishra, M., Kumar, V., Rani, R., et al. Tolerance and reduction of chromium (VI) by Bacillus sp. MNU16 isolated from contaminated coal mining soil. Front. Plant Sci. (2017). 8:778. doi: 10.3389/fpls.2017.00778PubMed Abstract | CrossRef Full Text | Google Scholar

Van De Camp K, Vernooij-Dassen M J F J, Grol R P T M, Bottema J A M How to conceptualize professionalism: a qualitative study Med Teach (2004);26(8):696-702. doi: 10.1080/01421590400019518. PMID: 15763872

Varma D, Meena RS, Kumar S, Kumar E Response of mungbean to NPK and lime under the conditions of Vindhyan Region of Uttar Pradesh. Leg Res (2017) 40(3):542–545

Wang L, Friesner R. A, and Berne B.J Competition of electrostatic and hydrophobic interactions between small hydrophobes and model enclosures Int J Biol Macromol. (2020)162: 853–865. doi: 10.1016/j.ijbiomac.2020.06.190 PMCID: PMC7308007 PMID: 32585269

Chapter 3

The imprint of social media on Zero Waste

Ritesh Pandey[1], Riya Srivastava[2] and Jyoti Prakash[1]*

[1] Amity Institute of Biotechnology, Amity University Uttar Pradesh, Lucknow Campus, Lucknow, India

[2] Kishan Lal Public College, Rewari-123401, Haryana, India

Abstract

Zero waste refers to "The conservation of all resources by means of responsible production, consumption, reuse, and recovery of products, packaging, and materials without burning and with no discharges to land, water, or air that threaten the environment or human health."(1) Zero waste is an important aspect of sustainability that influences the environment, economy as well as society. There are many approaches to achieve zero waste but most of them get lost in communiqué. Social media has emerged as an important platform to enforce commitment from various participants because of many reasons. The approach of social media and its influence on specific participants forms a commitment among the members of self-created groups which produces a higher impact on the targets to achieve zero waste. Social media is a useful tool for managing waste. With the right strategies, social media can be a low-cost tool to help spread knowledge and scope of waste management. Zero waste is not just a destination; it is also a set of guiding principles that work to get rid of trash at any and all points throughout the supply chain. The goal is to close the loop, redefining the entire concept of waste, ensuring resources are used for as long as possible before being returned to the earth with little to no environmental damage. This involves resource extraction, production, consumption, and management of discarded materials.

Keywords: Social media; environment; waste; pollution; zero waste management; sustainable development; conservation.

* Corresponding Author's Email: jprakash@lko.amity.edu

Introduction

Social media is proving to encourage the communities to become more responsible towards the environment. According to the research report, social media is used by 58.4% of people worldwide, 2 hours and 27 minutes is the typical daily usage. This makes it clear that any movement circulated in social media platforms can capture the eyes of millions of people. In addition to it, the zero-waste movement is now becoming widespread. The concept of zero waste was realized recently, as using the product till a time it can be used with intent to consume it to the end. Concept of zero waste goes strongly to support a sustainable planet by using the resources to their maximum utility. This paper has been structured to demonstrate the power of social media in educating an individual or a group of people about zero waste management.

What made social media so prominent?

Social media became so prominent because everyone wanted news and information at the earliest that is within no time. The previous modes of communication were newspapers where information could only arrive once during the day (in most cases), this was superseded by television where information was derived on real time but could be telecast with a small time lapse. It required a complete infrastructure for transmission of news. In fact, news transmission requires internal set up for broadcasting and news journalists capturing the event. Television transmission was superseded by internet news broadcast where a number of news channels also initiated broadcasting their news on the internet by creating their own web page but even this was limited to accessibility due to requirement of the presence of news journalists. Social media is connectivity between all the members registered on specific applications (such as Twitter, Facebook, WhatsApp or even through open platforms like YouTube). The messages can be transmitted by anyone who wants to transmit an event or information by posting it in the internet space. The individual as a member of the group can telecast the message at any point in time. The message could be received and used by as many users who received the message. The message can be re-transmitted and this phenomenon is known as a message getting viral. The support for making a message viral increased the number of messages, as anyone and everyone who was witnessed, telecasted the message in real-time. Boundaries were actually borderless. Global events could be telecast in real-time. Even those who are not journalists were making a mark in the relevant arena. There was no cost involved except for the regular costs of the internet. Today e-connectivity is a necessity, sometimes even more than the food which we consume.

What Is the Zero Waste movement?

The zero-waste movement is the collective pursuit of the zero waste principles, and as its popularity grows, individuals and communities around the world are helping push the agenda further. One of the most visible of many initiatives that are illuminating our consumption and waste generation habits is the 'Mason Jar Challenge' where participants reduce their waste down to a single mason jar over one year. However, the zero waste movement is gathering steam, and there are many more initiatives in the pipeline. In fact, people taking responsibility for their own waste production is only the tip of the iceberg and by increasing awareness around the issue, both politicians and producers will be forced to take more responsibility. Zero-waste stores, zero-waste agriculture, zero-waste fashion, and even entire zero-waste lifestyles and jurisdictions are now entering the mainstream, with more people than ever striving to meet zero waste principles.

Today, the Zero Waste International Alliance (ZWIA) positions itself as the only peer-reviewed authority on the topic, and its guiding principles are among the most comprehensive.(2) They ask the community:

- To adopt the ZWIA definition of the concept "Zero Waste: The conservation of all resources by means of responsible production, consumption, reuse, and recovery of products, packaging, and materials without burning and with no discharges to land, water, or air that threaten the environment or human health."(3)
- To establish benchmarks and a timeline for the zero-waste agenda to achieve objectives for tracking successes and assessing performance. In order to convince the public of this new course as soon as possible, communities should invest local resources and leadership in accomplishing meaningful and visible results within five years. Within ten to fifteen years following the adoption of a plan, several communities have set a target of diverting at least 90% of garbage generated from landfills and incinerators. Others have chosen longer time frames, such as the Urban Environmental Accords' target of zero waste generation by 2040. Establishing what is a realistic objective for your community while acknowledging the need of acting swiftly to combat climate change is a crucial step in the planning process.
- To engage the whole community – local, national, and international. It's crucial to avoid leaving Zero Waste in the hands of "waste specialists." The effort to achieve Zero Waste and sustainability requires the use of a wide range of talents. Each person has a responsibility. In order to involve all facets of the community in meetings, citizens or communities must take the initiative. To

achieve Zero Waste, all groups that provide waste reduction, takeback, reuse, recycling, and composting services should be included, including nonprofit organizations, grassroots movements, businesses, and governmental bodies. While their communities build longer-term policies and initiatives for the entire community, each of these groups and people should be pushed to seek Zero Waste at home, at school, at university, at work, and at play. Existing service providers must to be encouraged to accept Zero Waste as a goal and take advantage of the chances to minimize waste, offer takeback services to nearby manufacturers and shops, and aid communities and companies in becoming Zero Waste. Throughout the development and implementation phases of the Zero Waste strategy, continuous contact with all facets of the community should be maintained.

- To demand the management and conservation of all resources and not just the management of waste. No new incinerators may be constructed, and existing ones must be shut down. To stop all contamination of the air and water, landfill methods must be changed. This includes using Residual Separation and Research Facilities and pre-processing all residues at landfills before burial to stabilize the organic fraction and stop methane formation. However, it is not recommended to employ pre-processing facilities like this before disposing of waste in incinerators or other thermal treatment methods. Landfills are a significant source of both ground pollution and greenhouse gases, notably methane, which heats the climate 23–72 times faster than carbon dioxide. By converting resources that should be reduced or recovered into toxic ashes that need to be safely disposed of, incinerators and other burning and thermal treatment technologies like biomass burners, gasification, pyrolysis, plasma arc, cement kilns, and power plants using waste as fuel are a direct and indirect source of greenhouse gases to the atmosphere. Since every new incinerator will become unworkable during its lifespan due to the expense of replacing embedded energy and oil in products, neither landfills nor incinerators are a realistic solution to the problem of peak oil. By minimizing waste, reusing items, recycling, and composting instead of burning garbage or recovering landfill gases, more energy can be conserved and the effects of global warming can be lessened. Communities should oppose any attempt to build new incinerators, regardless of their appearance, and replace current landfills and incinerators with Zero Waste laws and initiatives, such as EPR, resource recovery parks, reuse, recycling, and composting facilities.
- Fund initiatives to educate and train Resource Managers to follow a Zero Waste strategy, to build programmes for handling community discards, to generate

green employment, and to enforce environmental laws using economic stimulus funding and fees assessed on tons of waste carried or landfilled.

- To perform zero waste assessments. To determine the quantity and kind of trash produced in a community, a waste audit should be conducted. If funding is not available, data can be gathered locally or from nearby areas that are similar. These audits should be used as a starting point to find potential for cost savings, cost reduction, and measuring the effectiveness of the reduction and rccovery programme. Determine which programmes and facilities for source reduction, take-back, reuse, recycling, and composting are required in order to make such services more user-friendly than mixed material collection and disposal services.
- To develop rules, laws, and incentives to promote zero waste. Communities have the power to greatly change the "economics" of their local markets by enacting new laws, regulations, and incentives. To move toward Zero Waste, communities should reorganize contracts and rules such that the expenses of collection and disposal are minimized.
- To remove government subsidies that allow and promote waste. Governments, notably in the US, have implemented a number of financial incentives to promote resource-wasting activities like mining and wood harvesting that are no longer necessary. Incinerators have also received government subsidies under the pretext of "Energy from Waste," despite the fact that these plants waste energy. Leachate, methane production, and continuous long-term care, which is an indirect subsidy for waste, have all been poorly addressed by government laws of landfills. Instead of using Pay As You Throw incentives, communities have chosen garbage rate systems that have made it cheaper to throw away than to recycle. Communities should demand the elimination of all of these additional subsidies as well as the subsidies for waste that are under their control.
- To expand zero waste infrastructure.
  - Zero Waste Infrastructure - Local governments and stakeholders should be involved in the creation of Resource Recovery Parks and other areas for reuse, recycling, and composting businesses to gather and process materials, create goods, and sell items to the general public.
  - Support zero waste initiatives at organizations and enterprises - Communities should mandate Zero Waste services for all enterprises and institutions, offer universal recycling and separate transportation services to all of them, and mandate source separation of waste materials to preserve the highest and best use of those materials.

- Constructing, Destroying, Clearing Land, and Remodeling (C&D) - Adopt citywide policies for demolition, reuse, and recycling (including requiring all contractors to submit plans and deposits to fulfil community objectives), and put the infrastructure and processes necessary to achieve Zero Waste. Work with green building initiatives to make deconstruction and reuse a priority and to make it a requirement for all new structures to provide room for recycling bins.
- Local businesses that are owned and operated locally - Communities should, if feasible, promote locally owned and managed community businesses to manage and utilize local waste products sustainably and to provide employment and training opportunities within the community.

What are the zero waste principles?

The zero waste principles include three underlying obligations that target different sections of society:

Producers' responsibility

In order to create a circular economy and a zero-waste future, producer responsibility is an essential policy tool. Producer responsibility transfers responsibility for post-consumer product and packaging management from local governments to producers.(4) This covers financial responsibility as well as day-to-day management of covered material. Since they have the greatest direct control over a product's design and production, manufacturers and brand owners can directly affect whether their goods and packaging can be recycled. However, producers typically have no vested interest in recycling programs performance. Programs for recycling are primarily administered and supported locally, either through user fees on households or funding from taxpayers. Local governments are burdened with rising costs to manage these materials yet have no control over how things are created or how simple they are to recycle. By transferring the expense of recycling to the manufacturers who produce the products, producer responsibility policies may help:

a) Provide locals access to more convenient recycling initiatives
b) More effectively recycle, decrease carbon emissions, and decrease plastic pollution
c) Encourage the development of ecologically friendly goods and packaging

Political responsibility

Zero waste is a crucial first step in the process of improving equity, environment, and other essential goals like health. Sustainable agricultural, economic, industrial, and community development are all related to zero waste. Every single individual on earth produces waste, making them a component of an unsustainable civilization. However, everyone may participate in the essential transformation toward a sustainable society with strong political leadership. In order to safeguard human health and the environment and to make the transition to a sustainable future, good political leadership in this area entails considering citizens as crucial allies. Governments must "govern" rather than "manage" the change to sustainable resource conservation methods. In order for individuals to assist communities in making the best decisions, there will be a major investment made in public outreach and education.(5)

Community responsibility

At the rear end of the problem, the community has responsibility for consumption, discard usage, and disposal.

Each represents a specific stage of the waste stream. Producers are at the front end, and they must take responsibility for product design and manufacturing. The Community sits at the back end, taking responsibility for consumption and disposal. In between, political responsibility must bridge the gap between community and producer, promoting both environmental and human health while enforcing new laws designed to promote the zero waste principles.(6)

The principles themselves are as follows, however, they are also constantly expanding to meet new challenges that arise as we continue to explore the realities of a zero-waste economy.(7)

- Design closed-loop systems
- Ensure processes (manufacturing, recycling, etc.) happen close to the source
- Conserve energy
- Don't export harmful waste
- Engage the community and promote change
- Keep products and materials in the loop as long as possible.
- Build systems that provide feedback for continuous improvement.
- Support local economies.
- Promote materials as resources.
- Minimize polluting discharges to land, water, and air.

- Consider the true costs of opportunities
- Promote the Precautionary Principle.[(8)]
- Promote the Polluter Pays Principle.[(9)]
- Develop adaptable, flexible, and resilient systems.

What caused an increase in waste?

People have the tendency to waste. The following reasons attribute to a general rise in the increased waste.[(10)]

Overpopulation

Due to increased industrialization and population, there is a significant amount of rubbish being created in many forms, including solid, liquid, sludge, and vapors. Every day, each city produces tones of solid waste from places including residences, hospitals, workplaces, and market centers.

These wastes might be harmful and toxic. The wastes are directly thrown onto city streets, garbage collection locations, etc. Where it subsequently accumulates and clogs the drainage systems, where the majority of it is still there.

Urbanization

Most cities are expanding quickly and occasionally beyond the boundaries that were originally intended because of the pressures of urbanization. A quarter of the city's population lives in unplanned regions, where there is typically no spatial information due to the lack of current maps.

Due to a lower population and a general lack of understanding, trash disposal earlier did not present a concern. However, a corresponding rise in the production of home and industrial solid waste was noted in big cities as a result of fast urbanization.

Some specific factors induced by business leading to increased waste are as follows:

An important phenomenon under business development is the short life of the product. The concept of new fashion and design has led to a shorter life of the product as consumers do not want the product to last for their life as was thought before. An example quite often quoted is the number of pairs of footwear used today by individuals is much higher than used half a century back and the same is true for clothes.

Consequences of the increased waste

Due to overpopulation and industrialization, every day, millions of pounds of plastic are made. And half of them are single-use goods like carrying bags, plastic straws, and food packaging. When consumers use these items, they discard them on the ground, where they eventually wind up in water bodies.[(11)]

Actually, the most common type of marine garbage is plastic debris. The fact that plastic doesn't break down is the main issue with it getting into the water. Instead, it is dispersed over great distances by ocean currents, having a huge impact on marine life.[(12)]

The ability of ocean plastic trash to inflict physical harm is its main drawback. It is conceivable that fish and other aquatic birds will sustain wounds and infections if they become entangled in plastic garbage. Keep in mind that marine animals need air to breathe. But when they become entangled, it becomes tough for them to access surface oxygen. As a result, they drown and perish.

Planktons are the main source of nutrition for many aquatic species. The sea creatures are more likely to suffocate, suffer harm, or pass away when they eat the plastic waste which may seem to be planktons but are not. Throughout addition, garbage floats in every ocean region. Thus, animals are more prone to consume it than any other food. Additionally, because it is difficult for marine life to digest plastic, their stomachs will remain full and unable to accommodate healthy food. The animals starve to death as a result of not receiving the nutrition they require.

Coral reefs are impacted by plastic pollution in the ocean. Marine life greatly benefits from coral reefs. They provide advantages like:

- Aids in fixing carbon and nitrogen
- Aid in the recycling of nutrients
- Provides a home to a variety of marine species.

In addition, numerous species breed and raise their young in the crevices that corals generate. However, when plastics get into the water, they help viruses spread. Additionally, the coral reefs are susceptible to contracting diseases when they come into contact with these creatures. Additionally, as plastic waste builds up in the ocean, it prevents sunshine and oxygen from reaching the corals. This prevents their growth, which has an impact on the marine life that depends on corals.

What can we do to clean up the plastic waste in the water bodies?

Using less single-use plastics will have the biggest impact. These goods play a significant role in the plastic pollution of water bodies. It is good to consider recyclable things such as carrying bags, metal straws, mugs, and plates. Also, you can participate in clean-ups of rivers and beaches to decrease the quantity of trash that enters the ocean. In addition to this, zero waste management is a responsible way to reduce plastic pollution.

Role of social media on zero waste

Following is some of the social media users who influence individuals to build a zero waste community. The examples have been taken from various social media influencers, who have made a difference to the philosophy of zero waste creation.[13] There are examples from global zero waste techniques compiled into one by Rogue Gone Vogue, Zero Waste China, Lauren and Anne-Marie Bonneau[14] who makes zero waste to come to the kitchen, Kathryn Kellogg and Gippsland Unwrapped are making a difference to plastic waste, Bea Johnson is looking at making a zero-waste effort at home.

Looking at the approach of social media and its influence on specific participant's forms we witness that there is a trend to connect in similar groups. After the proliferation of the usage of social media we find that there are groups with many similarities. People will connect in the same ages, connect through official groups, have less gender bias and will connect on common issues. Somewhere the groups have grown out of homes and neighbourhoods and connectivity is now global.

Waste management as it is, is the premise of the government or an organized structure. People would be more interested in an organized system which takes the responsibility of waste disposal. Looking at the number of units disposing waste it is a good idea to dispose of waste in mass. Such a waste disposal process has also given an understanding to civilians that waste can be created and if this post is at the right places then they fulfill their duty towards the nation and the globe. The process of waste disposal has become costlier with the increasing volume of waste and many times the intent is to dispose of the waste in manners unwanted and not sought for waste disposal. With this mindset it becomes very difficult even for the government to promote proper waste disposal systems among citizens. A very important part is to make the citizens participative. Just publicity and advertisement has not been able to bring in the results desired. Publicity and advertisements could only influence people at once but did not gain any kind of commitment. Social media as a means to propagate zero waste has gained more prominence and commitment from the citizens.

One of the frameworks developed by Shove, Pantzar and Watson in 2012 connects three elements: materials, meanings and competences. These three variables are used to look at insights on how individuals influence others with the help of online communities. An important observation has been made by

Hargreaves who has inferred that the practice itself is more important rather than the individual or the social structures which perform the practice. To make the concept of zero waste successful the practice of unwanted waste disposal has to be transformed into the practice of zero waste consumption. A behavioral change is required to turn this practice and this can be driven by creating an imprint on the mind to connect with the desired issue. Social media is repetitive and can be exhibited a number of times at the will of the user and there can be numerous messages which can create an imprint on the user. This gives credence to social media for creating an environment which promotes zero waste.

Social media is being used by propagators to offer tips on sustainable fashion purchases and use of products leading to a zero waste environment.

Ethical unicorn is a platform which has advised big organizations even like amazon.com to bring a difference in their workers pressures as the organization can switch to reusable packaging which makes it more manageable for the workers. Other issues handled are ocean plastics and plastic waste. The blog is visited by many.

The tourism industry has added enormously to waste and unsustainable practices. This happens because a tourist visits places and tries to move around on a shoestring weight. This drives the tourist to shed maximum weight anywhere and everywhere to be free from the burden. Instances have been noted where tourists have disposed of plastics, unwanted packing, excess food etc. Some of these instances figured on snow-clad mountains where disposal of dark colored wrapping generated heat absorption leading to melting of ice on the mountains. Normally such phenomena are well understood as a science student but in practice people don't realize as they did not get a chance to become a party to the same or have not seen such processes. Video clips exhibiting real damages can now be uploaded as and when the phenomenon is noted on the issue helps people to understand the phenomena. Many videos on the issue are available on social media. The social media which has been used for marketing is now simultaneously being used for influencing the consumer behavior and many pieces of advice on how to make sustainable prices are coming along with the marketing. Social media has been able to considerably reduce food waste in tourism. In fact, social media is now becoming a tool which is defining social norms and people are being influenced by what is trending on social media, they behave as a responsible consumer.

Many of the business outfits have now resorted to social media as a marketing tool to promote their products and services. One such example is of a restaurant in Germany which charges extra for any kind of food wasted by the customer. Examples like these were uncommon but through social media such videos and messages could travel around the world and the concept of being lavish by ordering additional food was challenged. The shift is prominent and food is not wasted as before. The leftovers are now requested for a bank and most of the time are either used for second film or used to feed someone else or even the pets thereby reducing wastage.

A NGO in Lucknow city which did not have huge resources could make a mark with the help of social media. This NGO used to prepare food for the attendants of patients in King George's medical College at cost. It was more as a service providing organization rather than a profit-making organization. It worked on a zero waste philosophy to make itself cost-effective. The organization grew tremendously and has now expanded to other hospitals in the city. There are others in use around the world who collect the leftover food from different restaurants and have organized distribution points where such food is kept in the evening. Whoever is interested can pull out the food for use. Such efforts promoting zero waste are possible with the presence of social media.

Flowers all across the globe are consumed in huge quantities. In India lots are consumed for worship more than for other purposes. Truckloads of flowers have to be dumped, which creates a lot of pollution. Flowers in India are also used to produce perfumes. The value chain of flowers spans to the perfumery industry as well as the manure industry. 'Help us Green' a social enterprise came up with the idea of converting flowers into perfumes and today it has grown into a $1 million Company. This idea has been fostered into a micro flower entrepreneurship franchise model. To propagate this idea word of mouth and social media have played a very important role.

A recent trend that emerged was to use plastic and thermocol plates for dining especially during the celebrations as they are considered as use and throw. Many times, it is witnessed that the waste plastic is set on fire, thereby increasing environmental pollution. Organizations understanding the need to save the environment and create a zero waste returned to the old concept of using plates made with leaves which are eco-friendly disposables. The process of production was designed so as to produce more presentable plates and spoons for the utility. Eco-friendly disposable utensils help to convert all the waste into organic manure. Such examples can easily be picked up from various videos on YouTube.

Today we have a high usage of plastic and we consider that plastic is a use and throw utility. As plastic is easy to produce and convert in any shape hence the utility of plastic increases in each and every sphere. Disposal of plastic became a big challenge. To create a zero waste utility out of the used plastic people started to come up with new innovations. Zero waste thinkers came up with using this plastic for the purpose of building roads which are three times the Strength of normal roads. Some plastic is now converted into shoes and shirts which are recycled products. Such ideas are now being propagated through YouTube.

Social media not only acts as a motivator to words best practices but it also is being used as a demotivator for unwanted practices. Celebrities are often exposed through campaigns for using high carbon footprint modes for travel even when they could have alighted a train.

The question how many things does one require to live is easily answered and exhibited through many intents and interests shared in the social media space.

While promoting zero waste efforts like distribution of jute bags for shopping instead of using plastic and polythene or the use of corn bags promotion is a requirement. Traditional efforts to promote such usages require campaigns and advertising through the traditional modes like television, newspaper and radio. All of these require a lot of effort and money and with social drivers the spender is only the government. Government promotion in developing countries like India requires tremendous allocations and efforts. Social media is a quick means of promotion and acceptance.

In fact, users of the internet and social media themselves witness competition from various societal pressures. Different groups would like to improve and excel in what they are intending to exhibit. Initially products were exhibited through marketing to enhance a consumerist culture. Now the trend has reversed and the channel is being exhibited to become more sustainable. Let us look at the drivers of the former and the latter. The former that is marketing of a product was driven by the companies to sell their products and they wanted to make more profit. Whereas in the latter case the interest is the survival of the planet and examples are driven more by personal interest of the user to have a cleaner and a beautiful planet than ever before. This involves lesser usages of products and lesser consumption, thereby leading to the concept of zero waste. It is not only the concept of who uses the least but it is now catching up with the concept of how to produce the least with longer usage. The competition in the social media space has driven people to become eco-friendlier than before, especially with an intent to gain more followers than the competitive groups.

Conclusion

Social media has influenced individuals and communities to make a step towards zero waste. The quantity of rubbish we generate in our world is a collective responsibility. We can realize the idea of zero waste and significantly improve the condition of the Earth if we use efficient waste management techniques. To reach zero waste, we must identify the problem, the location and extent of the problem, the method being used to solve the problem, and the potential consequences for the world if we continue to waste as much as we do. Zero waste is not only an ideal; it is attainable if we are all prepared to put out the effort. It would have been a herculean task to organize a complete governance structure and publicize the concept of zero waste in the absence of social media. The people at large scale have started to appreciate how they can employ ways and means to reduce waste and increase utility of products, if not hundred percent then a near to it. Social media is a self-starter and will continue to drive the philosophy of zero waste.

Disclaimer

None

References

Bailey, Kate. "Producer Responsibility." *Eco-Cycle*. www.ecocycle.org. Accessed August 21, 2022. https://www.ecocycle.org/zerowaste/overview/producer-responsibility#learn.

Clark, Duncan. "What Is the 'polluter Pays' Principle? | Environment | The Guardian." *The Guardian*. www.theguardian.com, July 2, 2012. https://www.theguardian.com/environment/2012/jul/02/polluter-pays-climate-change.

Drell, Cady, and @marieclaireus. "Is It Possible to Live Trash-Free? An Experiment in Zero Waste." *Marie Claire Magazine*. www.marieclaire.com, July 30, 2019. https://www.marieclaire.com/culture/a28468755/is-the-zero-waste-movement-doable/.

Epstein, Charlotte. "Precautionary Principle". *Encyclopedia Britannica*, 28 Feb. 2019, https://www.britannica.com/topic/precautionary-principle. Accessed 21 August 2022.

Ethical Choice | For Sustainable Future. "How Plastic Pollution *Affects Marine Life*." myethicalchoice.com, October 28, 2020. https://myethicalchoice.com/en/journal/pollution/plastic-pollution-affects-marine-life/.

Global Recycle. "What Are the Effects of Increasing Waste? | Updated August 2022." globalrecycle.net, February 12, 2022. https://globalrecycle.net/what-are-the-effects-of-increasing-waste/.

poster. "What Are the Reasons to Increase the Amount of Waste." c.worldmisc.com. Accessed August 21, 2022. https://c.worldmisc.com/read/what-are-the-reasons-to-increase-the-amount-of-waste.

Zero Waste International Alliance. "Zero Waste Community Principles - *Zero Waste International Alliance*." zwia.org, June 24, 2018. https://zwia.org/zero-waste-community-principles/.

Zero Waste International Alliance. "Zero Waste Community Principles - *Zero Waste International Alliance*." zwia.org, June 24, 2018. https://zwia.org/zero-waste-community-principles/.

Zero Waste International Alliance. "Zero Waste Definition - *Zero Waste International Alliance*." zwia.org, June 24, 2018. https://zwia.org/zero-waste-definition/.

Zero Waste. "The Benefits of Zero Waste | *Zerowaste.Com*." www.zerowaste.com, January 13, 2021. https://www.zerowaste.com/blog/the-benefits-of-zero-waste/.

Zero Waste. "What Is Zero Waste? Learn About the Movement & Why It's Important | *Zero Waste*." www.zerowaste.com, July 23, 2020. https://www.zerowaste.com/blog/what-is-zero-waste-a-guide-to-resource-recovery-and-conservation/.
Zero Waste. "What Is Zero Waste? Learn About the Movement & Why It's Important | *Zero Waste*." www.zerowaste.com, July 23, 2020. https://www.zerowaste.com/blog/what-is-zero-waste-a-guide-to-resource-recovery-and-conservation/.
Zero-Waste Chef. "*Zero-Waste Chef*." zerowastechef.com, September 30, 2020. https://zerowastechef.com/.

Chapter 4

Approaches Towards Waste Recycling and Zero Waste for Sustainable Waste Management

Arjun Singh Kohli and Rachna Chaturvedi*

Amity Institute of Biotechnology, Amity University Uttar Pradesh, Lucknow Campus, Lucknow, India

Abstract

Waste management includes several objectives and perspectives ranging from a reduction in waste generation and production to efficiently managing the generated waste. For achieving the set principles of an effective waste management system, the methodologies adopted must be aimed at being sustainable, economic, and user-friendly which consider involving every section of the society linked to the waste ranging from the waste-generating body to the waste-handling body. This chapter discusses various approaches that aim towards a sustainable management of the world waste particularly via recycling the waste that is generated that will involve a re-introduction of the produced waste into the system for a circular economy and securing zero waste production milestone to ensure that the source of waste is affected and hence, to work with the capacity that only that much waste is introduced as much can be efficiently handled or eliminated. Zero waste is the universal idea of waste management which acknowledges waste as a source created in the temporary period of the procedure of source utilization. Zero waste approaches might be functional for enterprises, neighbourhoods, manufacturing sectors, institutes, and residential homes for sustainable development.

Key words: Waste management, Recycling of waste, Zero waste, sustainable development

* Corresponding Author's Email: rchaturvedi1@lko.amity.edu

Introduction

Sustainable Waste Management

When a question regarding the role played by sustainable waste management in achieving a broad sense of circular economy is tossed around the table, it is identified and understood as a defined systemic approach of achieving an overall economic development by mainly unlinking growth from the utilization of limited natural or unnatural resources, and all of this within the framework of an environment-friendly attitude. Sustainable waste management not only helps in handling the problems generated by a linear consumption society but also provides with solutions which are easy to implement against the issues of waste handling and management. Any waste management system which facilitates the sustainable execution of policies and procedures aims to part ways with traditional practices of "flaming, flushing or flinging" waste and moves towards a much more flexible, adaptable and sophisticated approach (Jeffrey K. Seadon, 2010).

Waste Reusing and Recycling

Sometimes it seems that a product could not be consumed then product should reusable and recyclable. Reusing is favoured to alternatives grading choices lower down the hierarchy since it cannot be processed without processing of other new resources, which involves cost, power, and different sources. Reuse, one of the most important views of the zero-waste viewpoint.

After reusing next option is recycling. This is where management starts to look like traditional waste management here, we are now processing with resources that have achieved the end of their functional life in their present form. Recycling, like reusing, keeps resources in the loop, dodging the necessity to remove pure form as well of the negative influences of just discarding waste. Recycling is considered less desirable than the previous options since it requires energy, money, and resources to turn waste back into usable materials.

Waste recycling discusses the many benefits of waste utilisation and re-utilisation at its very best and ultimately reduces the loads of waste reaching landfills or incineration centres. The basic process of collecting and processing recyclable waste and transforming it into new products directly impacts the economy as there is a reduction in the energy consumption observed in both eliminating waste and making a new product.

Zero waste

On the other hand, zero waste can be visualised as a significant contribution towards waste recycling as a zero-waste management system aims at reusing all the waste that is generated and ensuring a sustainably enriched waste management protocol. In 1973, one of the scientist Dr. Paul Palmer initially used the term "zero waste" to recuperate compounds (Palmer, 2004). Zero Waste (ZW) is described as "a scheme to propose and manage the products and procedures for elimination of waste, holding of materials and restoration, not burning or burying" (Zero Waste International Alliance, ZWIA, 2004). Zero waste method is one of the maximum broadly considered hypothesis but still the greatest debateable subject in current era in system of waste management study (LaBrecque et al., 2015).

Acknowledgement of The Challenges Encountered and the Approaches Associated with Waste Disposal and Waste Management Strategies (Sunil Kumar et al. 2017)

Growth of Mega Cities

Megacities are a comparatively current scenario, connected with globalisation of the basis of wealth, civilisation, and expertise (ISWA [International Solid Waste Association] 2012.) With the historic growth in population and the rapid urbanisation that brings exhaustive utilisation of the available resources and commodities also comes the growth of waste that is generated alongside. The largest hurdle in the failure of many or most of the waste management strategies is the exponential production of waste. As per a late-2021 World Bank report, 2.12 billion tons of waste is dumped annually round the globe attributing to the massive waste problem. By the year 2050, the world will have been dumping 3.4 billion metric tons of waste every single year into its landfills. These statistical datasets not only reveal the very grave atrocity of the waste we generate faced by the environment but also highlights the unsustainable linear consumerism that is being practised by populations globally leading to piling up of waste and only minimal percentages of that waste making it to the recycling centres. The growth in globalisation, rapid technological developments, and the use-and-throw approach visible in all industries from fashion to food, electronic to vehicular is single-handedly producing large amounts of waste that not only faces inadequate handling strategies but also is mishandled due to lack of knowledge among masses about waste handling.

Infrastructural advancements and Revenue Shortcomings

A well-maintained and fulfilling civil infrastructure plays a pivotal role in meeting public demands and achieving sustainable development goals simultaneously. A high-quality state-of-the-art waste management system can cater to almost every section of the society and contribute largely to the economic growth of a nation. All the waste that is rejected and eliminated as useless actually is a potential resource which harbours resources in the form of energy, materials and nutrients Only effective waste management can ensure efficient resource extraction from waste reducing the amounts reaching landfills or open burning centres. The conversion from litters to usable form can only be attained by investment in SWM as this varies on a organized by collection of activities to improve marketplaces and boost retrieval of reusable/recyclable items (Sridevi P, et al. 2012) The governments need to invest more into these projects for an enhancement of waste management programs as a result of which development of markets can help in maximal recovery of recyclable substances from the waste that is produced. Existing technologies, well-coordinated frameworks and local waste segregation and management bodies can be unified to produce advantageous outcomes

Involvement of the Informal Waste Management Sector

The informal sector is categorized by minor-scale, employment-demanding, principally unregulated and unregistered low-technology manufacture or supply of resources and public services (Wilson D.C, et al. 2006). The transition towards a much more sustainable society requires a lot more sophistication, awareness and integration into the waste management systems. The locally functioning informal sectors are composed of small-scale, labour-intensive, unregulated, and unregistered low-technology manufacturing or circulation practices. These practices serve employment to the unemployed and is characteristically actively functional but has several demerits involved These local systems are unintegrated and work at individually governed levels. Many of the waste-pickers are unaware of the categories of waste they are exposed to and handle regularly. Lack of knowledge, technological calibre and unhygienic conditions cloud the many benefits of such a system. These local bodies must be integrated into the central or formal sustainable waste management program to ensure efficient regulation, modulation and patterned execution of policies to achieve waste management.

Lack of Training

Untrained and unskilled labour contributes to an ineffective regulation of procedures. Workers are unaware of the various types of waste and how to handle them. A proper training in the field of waste collection, transport, segregation, and disposal can help in a stipulated growth in the waste reduction efficiency. Workers without basic training might avoid taking the necessary precautions to be taken care of while handling hazardous waste types such as radioactive, medical and laboratory waste. These workers might only ensure regular collection of waste but all the waste, after collection might end up in the same landfill where no segregation is done, and this opens door to mishandling of waste and several ecological issues.

Lack of Accountability

The general public must work in unison with the governing bodies to ensure that minimal waste is generated in the first place and that there is efficient reduction in the amount of waste making it to landfills and dumping grounds Approximating the amount and appearances of Municipal solid waste in India and prediction forthcoming waste production is important to effective waste supervision arrangement (Rana P.R, et al. 2014) People must align themselves to a sustainable lifestyle as per which most of the waste is either reused or recycled and many new avenues of environment-friendly alternatives can also be invested. To adopting an economic and sustainable lifestyle, waste management systems can be unburdened of the lots of waste that could have been re-utilised after a basic transformation and other ecological and environmental distresses can thus be averted.

Lack of Strategic and Environmental Awareness

Another hurdle is seen in terms of approaches which lack strategic spontaneity or effectiveness. Such approaches serve short-term benefits and account for minimum to no benefits at all. It is necessary that all policies, rules and regulations and interwoven and coordinated under a regulatory governing body which works in accordance with the stipulated sustainable development goals. This would allow for a monumental transformation accompanied with historic public involvement. Lack of environmental awareness can be attributed to several fallacies in the system itself. Masses have been reported saying that they are environmentally sensitive and aware but an ineffective system doesn't allow for transformation. If the authorised bodies ensure rightful management, the public and other groups will follow

Sustainable Waste Management: Policies and Measures Including Hierarchical Placement of Waste

Sustainable waste management must work as a coordinated system within a regulated framework of policy execution and general mass awareness. Any or all of the policies that form a part of this system must be designed in accordance with a method of achieving them and executing them at both the public and the state level for an efficient and systematic approach towards waste elimination (Lee, R.P. et al. 2020)

Waste can be placed on a hierarchy as per which a focus on avoidance and reduction, reuse, recycling, recovery of energy and treatment or disposal is laid. This method helps in prioritizing actions where the most commendable use of resources with minimal waste generation is ensured. As per the hierarchical pyramid, practices with possible energy recovery and less amounts of waste being produced are considered as the top of the pyramid. These principles highlight as to how this hierarchical arrangement of waste management practices is pivotal in ensuring sustainable handling of waste.

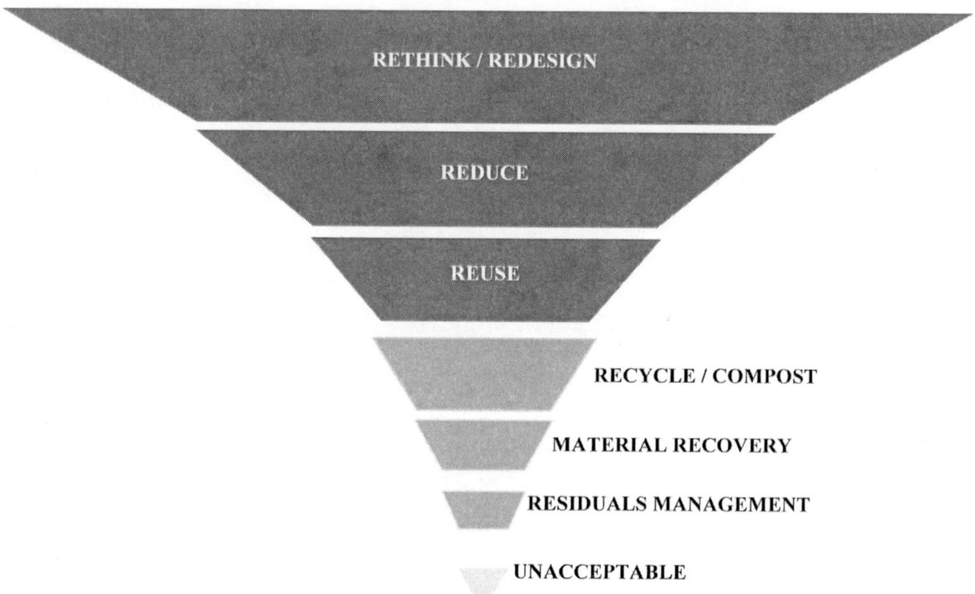

Figure 1. *Hierarchical Arrangement of Waste; Pyramid of Waste Management Hierarchy.*

Rethinking and Redesigning

Rethinking and redesigning involve the adoption of a flexible, proliferative and transformative approach towards the items and products that form an essential part of our daily lives and the excessive amounts of waste associated with them. Adopting a newer and much more adaptable way of expenditure, conscious and sensitive investments and a general awareness towards a variety of factors such as the production cycle of various products, carbon footprints of the manufacturer and our own self, packaging, pre-consumption and post-consumption waste can help in assuming a better outlook towards a sustainable waste management strategy (Lee, R.P. et al. 2020) This forms the base of the downward pyramid of waste hierarchy leading its way into the much narrower perspectives of waste management as we move down.(Fig 1)

Avoidance and Reduction

A reduction in consumption and an improvement in the efficiency of the substances, materials and products that form a central part of our lifestyle can serve as an outlook for avoidance. Avoiding practices that contribute to large waste production must be prioritised by individuals and businesses alike. Looking out for more sustainable alternatives to use and products with minimal packaging can be the first step in framing a habitual pattern of waste avoidance and reduction. Businesses and upcoming entrepreneurial ventures must ensure the produced products utilise the fewest resources for an outcome and that any single-use item must be avoided.

Reduction in the amount of produced waste can directly impact the process of sustainable waste management by ensuring that only minimal amount of waste is left to treat or dispose of. Countries like China have imposed a ban on amounts of imported solid waste which includes up to 32 categories of waste. These imports are performed to facilitate compensation for the shortage of adequate domestic resources. Post ban on imported waste, the NDRC (National Development and Reform Commission) of China aims to focus on policies such as reducing waste generation in the country and charging citizens for waste generation. As per a new regulation, an effective measure to hold waste producers responsible for their waste generation can be legalised.

This practice stresses its applicability on substances that form immediate waste i.e., these items are rendered as waste quickly after discard. If an alternative for single-use goods or disposable items can be found, the amount of linear waste can be drastically reduced.

Reuse and Recycling

In scenarios where avoidance and reduction are inevitable or unavoidable, the focus must be shifted towards items that can be reused or recycled. Reusing items can be significant if the items can be upscaled to be used for a different activity or be donated to someone who can wield the usage of the product without the expenditure of money, resources and energy that would come along the production of a new product. This practice also plays a central role in zero-waste strategy as hardly any or all of the items are discarded as waste and automatically the previous point in the pyramid i.e., reduction of waste is also honoured.

On the other hand, recycling has a much different outlook in the entirety of the waste management practice as this requires energy, resources, and money to facilitate a transformation of the waste product into a new item because recycling is particularly performed for articles that, in their current form, have reached an end of usable life (Thiounn, T., and Smith, R.C. 2020). Though recycling is considered undesirable, it keeps the materials in the loop, eliminating the need of extracting virgin resources. For instance, aluminium recycling saves up to 90% of the energy which is otherwise expended in aluminium production using virgin metal.

Energy Recovery

Further up in the pyramid, conversion or transformation of waste into usable energy such as heat, electricity or fuel can be achieved via multiple practices such as that of incineration, gasification, pyritization, LFG (Landfill Gas) recovery and anaerobic digestion.

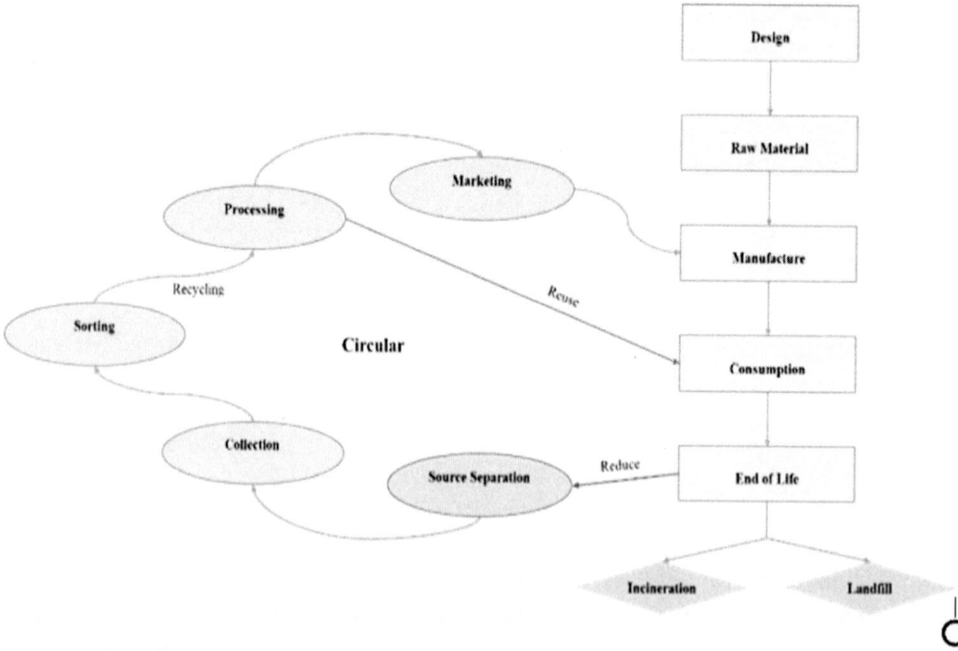

Figure 2. *Cyclic and Linear Resource Flow.*

Besides an attempt to reduce the amount of physical waste reaching the landfills and piling up in localities, combustion is extremely commonly used as a method for energy recovery. This practice of recovery of energy from the burning of Municipal Solid Waste (MSW) reduces the dependence on fossil fuels for energy production and supply. Analysing the aerosols and fumes that are released upon combustion of waste and the non-compliance of the practice with zero-waste strategy, energy recovery has been viewed as an undesirable rather, unacceptable compromise. (Fig 2)

Treatment

This step forms the tip of the pyramid and is seen as the last resort or the most undesirable step in the process of efficient waste management. It involves the dumping of waste in landfills and performing incineration sans energy recovery. Treatment methodologies must be minimally performed as these strategies will always be applicable to some amounts of waste that form the remainder of the entire volume after passing through the previous stages of a sustainable waste management approach. Hence, disposal must be viewed as a last option and not as the only option.

Approaches Towards Sustainable Management of Waste

Adoption of practices which are simpler to execute both by businesses and self can help in an extremely effective implementation of successful waste management.

NO to Single-Use Items

Scrapping single-use items or replacing items limited to use once with reusable goods can reverse multiple inefficiencies encountered in the global waste management scenario. All of the single-use items end up in a landfill increasing the load on public and private sectors to treat them. Most or all such items are easy to produce, supply and handle in many forms such as plastic which has made it to almost every single sector of industrial and commercial front. But, as waste, these commodities are generally non-biodegradable or non-perishable and due to their one-time usage, they form a huge percentage of the solid waste chugged out of municipalities, households, industrial and economic sectors. Looking for and choosing alternatives for such items can reduce the workload, capital and infrastructure that is poured in treatment and disposal of these products. Moreover, such a switch allows for a transcendental approach towards a sustainable lifestyle and hence a sustainable society.

Digital Switch

A strategy with intricate demands, this is applicable to fronts which can accommodate and allow for a transition and feasibility factor. Paper and paper products alone form the largest part of MSW due to their universality, easy accessibility, and perusal. Such a switch can be facilitated in businesses and economies where a transition of this sort is both possible and feasible. Paperless transactions, documents, invoices, digital invitations, and memos can bring about a transformational achievement in cutting down the amounts of waste reaching landfills.

Trash Alternatives

It is necessary to understand, comprehend and imply that only after availability of and accessibility to certain specific means can sustainable living be adopted. It is essential that businesses and governments work alike in providing means and resources to their employees and the public respectively in order to ensure action is taken. This can include activities as simple as providing separate bins for different waste types, going paperless, making available services that cater to such requirements and needs and feeding general awareness in the direction of a sustainable approach. All these approaches would single-handedly contribute to a conscious and effective management of waste.

Reusing Items via Donations

Another aspect of a sustainable approach of waste management is in understanding the usability of a particular item or product. The act of donation simply implies that the item no longer of yourself use can be fit for usage and utility by someone else. These items could include overstocked food grains, leftover food products, out-of-promotion items, old hardware, and clothing.

Sustainability in waste management can be ensured anywhere provided, a necessary framework is in effect to obtain significant involvement from governments and masses alike. Any amounts of waste, if left unchecked, can account for problems large enough to not consider (Menikpura, S. N. M. et al. 2012 and Sungmin K. et al. 2020.)

Principles of Zero Waste: Steps Towards Smarter Waste Handling

The definition of zero waste as per ZWIA (Zero Waste International Alliance) is: "Zero Waste: The conservation of all resources by means of responsible production, consumption, reuse, and recovery of products, packaging, and materials without burning and with no discharges to land, water, or air that threaten the environment or human health".

Zero waste strategy is all about encouraging a much more circular approach in terms of the usage of our resources. It is about reducing the environmental impact of our disposable culture and pushing economies towards not sending waste to landfills, incinerators or the ocean. It has been identified as a certain set of principles that must be followed throughout the entire chain of waste production from resource extraction to production, from consumption to energy recovery and disposal and that the loop is closed i.e., the resources must be used to their possible fullest before being re-added to the earth with little or no impact. These principles include:

Designing Closed-Loop Systems

Zero-waste project is widespread as an ecological production approach. Ensuring maximum energy utilisation and product utility extraction before the product is rendered as waste and sent for disposal in a landfill or an incineration centre. This forms a major part of the larger objective of a circular approach that is contrary to a much environmentally costly linear method of lifestyle, and it keeps the elements being used in check until the maximum possible usability is extracted. Reusing objects and recycling them to obtain an upscaled and refurbished product can fulfil the narrative of a closed-loop system.

Energy Conservation

Envisioning a product manufacturing line to focus on minimal or extremely low energy consumption while producing a product can help in keeping in line the fact that the product produced is environmentally friendly and upon disposal many of its energy can hence be recovered. This way every product can have a reduced environmental impact before being made accessible to in the market.

Avoiding the Export of Harmful Waste

Ensuring that the waste that is produced is handled close to its source of generation to avoid unnecessary impacts on transportation costs and burdening of local waste management sectors. If harmful waste is exported to another land, this will lead to uncalled for environmental impacts that are accountable for several health risk factors and waste accumulation in the country that has imported this waste.

Building Systems that Provide Feedback for Continuous Improvement

A sustainable waste management system is of an extremely transformative and flexible nature. A zero-waste strategy which calls for public involvement and participation in terms of both physical presence and feedbacks can work efficiently to bring to fruition the set goals. These feedback loops, once incorporated, can bring forth multiple opportunities for improvement such as several alternatives, ideas and innovations that focus on waste reduction and waste handling. Such flexibility and adaptability will ensure government action and revolutionising transformations in terms of waste handling towards a sustainable economy.

Minimising Polluting Discharges into Land, Water or Air

Ensuring that any or no amounts of waste is subjected to open dumping, dumping into the sea or open burning. These activities are of an ignorant approach towards finding alternatives for a better and efficient waste management system. Inadequate thinking and ignorant attitude cannot redesign the waste management systems. To make a society sustainable, adaptable and resourceful ideas are necessary, with treatment and disposal only serving as the last option left.

Promoting the 'Precautionary Principle'

Ensuring the promotion of an approach that focusses on the adoption of preventive measures in terms of addressing risks associated with the public or the environment. This principle forms a strict basis behind the laws that stipulate certain specifications in terms of the regulation of the generation, transportation, treatment, storage, and disposal of hazardous waste.

Promoting the 'Polluter Pays Principle'

Ensuring that the ones responsible for waste production are solely responsible to manage the produced waste to avert any damages to public or environmental health. With such a principle in effect, the waste management strategies can be applied globally by holding industrial giants responsible for their carbon footprints, emissions, and waste generation.

Developing an Adaptable, Flexible and Resilient System

A system which highlights the role of everyone involved and ensures efficient policy execution is easy to follow. Such systems allow for holistic improvement in terms of maintaining a sustainable lifestyle. A system which can easily adapt to the current scenarios and can respond to public waste handling needs is required to be functional at the core of sustainable waste management. The outlook of people towards global issues has been changing with people adopting sustainable alternatives and ensuring household-level consciousness in terms of waste management. (Hamid Baba, S, et al. 2020). The development of a waste management system that caters to such activities can turn out to be pivotal in moving towards a zero-waste society.

Cradle-To-Cradle Thinking: The Concept of Zero Waste

The cradle-to-cradle thinking, or strategy has been defined as a circular model that focusses on waste minimisation and keeping resources in the loop for as long as possible. This is the definition of a closed loop system that works in the direction of a sustainable approach by bringing to centre the practices of reduce, reuse, and recycle. This approach is contrary to the approach of cradle-to-grave which is a linear model and an open system.

Organic farming and the associated natural cycles can be cited as perfect incidences for cradle-to-cradle strategy. As per this, food is grown via natural methods without the involvement of harmful pesticides, synthetic chemicals or herbicides and is distributed and consumed via carbon-neutral distribution channels and compostable or reusable packaging. Post consumption, the food waste generated is composted, thus, closing the loop as this compost contributes to the growth of more food (Awasthi, A.K. et al. 2021)

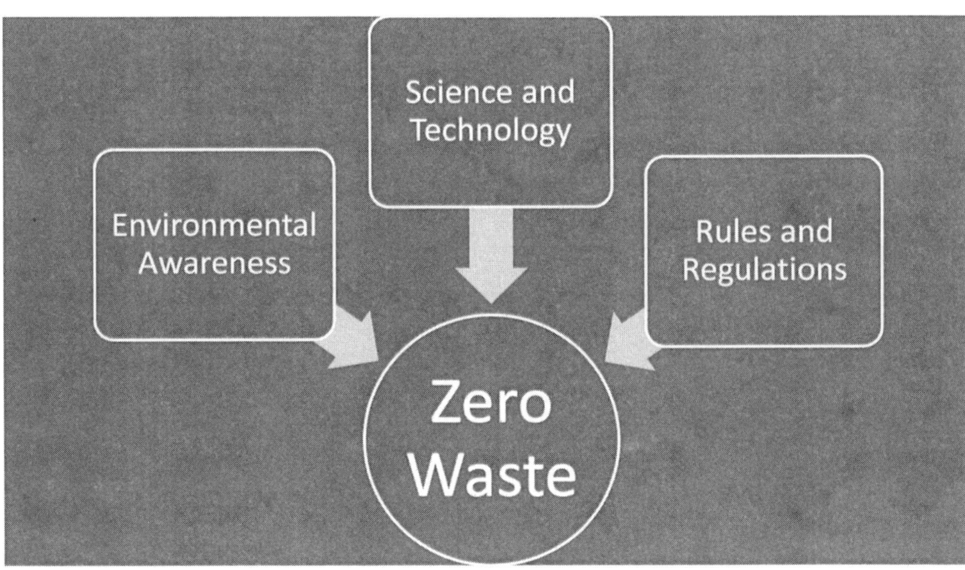

Figure 3. *Strategic approach for zero waste*

Cradle-to-cradle thinking can be efficiently applied to several procedural programs to ensure zero waste and hence help in pursuing a sustainable approach in waste management practices. A zero waste hierarchy aligns itself with the common waste placement pyramid of reduction, reuse, recycling and rethinking, redesigning and treatment (Menikpura, S. N. M. et al. 2012)

Zero waste proposal is having three main strategies as involvement of Science and technology, by dissemination for environmental awareness and strict compliance of rules and regulations for biomedical waste disposal (Fig 3)

Conclusion

A sustainable waste management system is supposed to be reliant on the system's strength and not its expansion, its processes, products and procedures and leverage points that are essential in effecting system change. Such systems must be flexible, adaptable, and resilient and must be fed with both positive and negative feedback loops, government capital investment, strong infrastructure and public involvement. Such a system can help achieve recycling and zero-waste goals Illegal dumping activities, open combustion of waste and informal treatment strategies highlight inadequacies in thinking and management. Zero waste, on the other hand, aims to promote the redevelopment and reestablishment of life cycles associated with resources and to ensure the reuse of all the products. Pivotal transformations have taken place amidst a stir in the societies and economies regarding the threat of a climate crisis dawning upon and significant projects and strategies are lined up ahead, indicating a gradual shift towards a sustainable outlook towards lifestyle, expenditure, and utilisation of resources.

Cities and localities are working alike in terms of waste management which will require more awareness and active involvement of conscious planning and execution specifically in developing countries to allow space for a zero waste-generating society. A change in thinking besides a change in general outlook and methodologies is necessary and can evoke a sense of acknowledgement towards the waste that forms an undesirable yet equally omnipresent factor of our land and lifestyles.

Disclaimer

None

References

Awasthi, A,K, Shankar Cheela, V.R. D'Adamo, I, Iacovidou, E, Rafizul Islam, M, Johnson, M, Reed Miller, T, Parajuly K, Parchomenku, A, Radhakrishan, L, Zhao, M, Zhang, C, Li J, Zero Waste Approach Towards a Sustainable Waste Management. Resources, Environment and Sustainability 3 (2021) 100014 https://doi.org/10.1016/j.resenv.2021.100014

Group. Retrieved from http://www.zwia.org/main/index.php?option=com_content&view=article&id

Hamid Baba, S, Skinder B. M. Bhat M.A. Zero Waste: A Sustainable Approach for Waste M management. Innovative Waste Management Technologies for Sustainable Development 134-155 (2020).

ISWA (International Solid Waste Association). 2012 Globalization and waste management final report from the ISWA task force. See http://www.iswa.org/knowledgebase/tfgfinal (accessed 27 June 2015). Google Scholar

Jeffrey K. S. Sustainable Waste Management Systems. Journal of Cleaner Production

La Brecque E, Curtice C, Harrison J, Halpin P. N, Parijs S.V, Biologically Important Areas for Cetaceans Within U.S. Waters – East Coast Region (2015) Aquatic Mammals 41(1): 17-29 DOI: 10.1578/AM.41.1.2015.17

Lee, R, P, Meyer, B, Huang, Q, Voss R. Sustainable Waste Management for Zero Waste Cities in China: Potential, Challenges and Opportunities. *Clean Energy*, Volume 4, Issue 3, (2020). Pages 169–201, https://doi.org/10.1093/ce/zkaa013

Menikpura, S. N. M, Gheewala, S. H, Bonnet, S, Chiemchaisri C, Evaluation of the Effect of Recycling on Sustainability of Municipal Slid Waste Management in Thailand. (2012). Waste and Biomass Valorization 4(2):237–257 DOI: 10.1007/s12649-012-9119-5

Palmer, P. (2004). *Getting to Zero Waste*. Sebastopol, CA: Purple Sky Press.

Rana P.R, Yadav D, Ayub S, Siddiqui A.A. Status and challenges in solid waste management: a case study of Aligarh city. J. Civil Eng. Environ. Technol. 1, 19–24.(2014) Google Scholar

Sridevi, P., Modi, M., Lakshmi, M. and Kesavarao, L. A Review on Integrated Solid Waste Management. Int. J. Eng. Sci. Adv. Technol, 2, 1491-1499. Sridevi P, Modi M, Lakshmi MVVC, Kesavarao L. 2012A review on integrated solid waste management. Int. J. Eng. Sci. Adv. Technol. 2, 1491–1499. (2012) Google Scholar

Sungmin K, Kim, W, Shin, S.-C, Shin, J.. The Economic Value of Sustainable Recycling and Waste Management Policies: The Case of a Waste Management Crisis in South Korea. Waste Management Volume 104, (2020)., Pages 220-227 https://doi.org/10.1016/j.wasman.2020.01.020

Sunil K, Stephen R. S, Geoff F, Costas V. S, Jyoti K, Arya S.R, Kumar R, and Cheeseman C. Challenges and Opportunities Associated with Waste Management in India. (2017). https://doi.org/10.1098/rsos.160764

Thiounn, T,. Smith R.C, Advances and Approaches for Chemical Recycling of Plastic Waste. Journal of Polymer Science Pages: 1347-1364 Volume 58, Issue 10 (2020). https://doi.org/10.1002/pol.20200296

Volume 18, Issues 16–17, (2010). Pages 1639-1651

Wilson D.C, Velis C, Cheeseman C. Role of informal sector recycling in waste management in developing countries. Habitat Int. 30, 797–808. (doi:10.1016/j.habitatint.2005.09.005) Crossref, ISI, (2006) Google Scholar

Zero Waste International Alliance. (2004). Zero Waste Definition Adopted by Zero Waste Planning

Chapter 5

Phytoremediation: An Eco-Friendly Tool for Industrial Waste Management

Arjun Singh Kohli and Jyoti Prakash*
[1] Amity Institute of Biotechnology, Amity University Uttar Pradesh, Lucknow Campus, Lucknow, India

Abstract

Over the course of the past two decades, extensive globalisation, rapid urbanisation and industrialisation and the increasing population has resulted in the generation of large amounts of waste which is toxic and hazardous in nature. With an approach towards increasing the agricultural produce, farmers and industrial forefronts have worked alike in utilising huge quantities of synthetic materials and chemical subjects which are ultimately discharged into the mainstream ecosystem resulting in widespread toxication and pollution of the natural resources. The propping up of vitally efficient new industries of cosmetic materials, synthetic products, disinfectants, additives, analgesics, plasticisers and phthalates has enormously and single-handedly contributed to the exponential increase in the amounts of newer and harder-to-handle pollutants reaching the waste management chain. Phytoremediation has very recently emerged as a green technology which wields the many benefits of bioremediation to use a variety and range of plants to perform the remediation of soil, sediments, groundwater environmental contaminants and surface water with toxic organics, metals and radionuclides. Strategies of this sort have been eyed as successful and technologically efficient in terms of achieving a sustainable developmental approach towards waste management. This chapter discusses the several approaches performed using phytoremediation as an eco-friendly tool and the many benefits that are reaped via such practices.

Keywords: Phytoremediation, bioremediation, pollutants, waste management, Toxic Contaminants

---

* Corresponding Author's Email: jprakash@lko.amity.edu

Introduction

Phytoremediation offers solutions for the remediation of hazardous pollutants that are particularly and very significantly released from industries and the many commercial sectors that dump excessive waste into the land, water and air. These solutions, compared to conventional and traditional methods, are cost-effective, aesthetic, long-lasting and sustainable. Phytoremediation is the combination of the terms – "phyto" meaning plants and "remedium" meaning to clear and restore and encompasses the age-old idea of utilising plants for remediation practices. These practices are efficient technologies that focus on the removal, transfer, stabilization and/or degradation of pollutants and contaminants found in soil, sediments and water. Analysed and practised as a significantly effective, efficient and beautiful (in terms of aestheticism) technology, phytoremediation has been alternatively termed as green remediation, agro-remediation and vegetative remediation and makes use of photosynthetic plants including shrubs, herbs and trees for the activity of restoring and reclaiming of contaminated environments. Regarding a discussion on the many strategies and magnificent abilities plants utilised for remediation and reclamation strategies possess, it has been elaborated as to how plants make way for the habitation of a broad spectrum of microorganisms via roots, leaves and stems and these microbes further enhance the treatment processes which include a variety of biophysical and biochemical practices like adsorption, hyperaccumulation, transformation, translocation, transport and mineralisation.

Heavy Metals – An Introductory Background

Upon elaboration, the term heavy metals can be defined as any transitional metal with an atomic mass up to 20 and specific gravity above 5. These metals form the large portion of inorganic pollutants in the ecosystem and after an estimate, large areas of land and water have been found to be contaminated with these contaminants (Chandra R. et al., 2017). This can be attributed entirely to the profound involvement of the industrial sector in the field of waste production and release via the several forefronts of crude substrate extraction, substrate refining, production, manufacturing, processing, reforming, finishing and ultimately the retail of the product. The extensive utilisation of chemicals, fertilisers, synthetics, agrochemicals, complex compounds for product production, refining and enhancement contribute to many sour aspects of the industrial niche (Chandra R. and Kumar V., 2017). Activities like improper treatment of wastewater or effluents before release into water bodies, open dumping of products of toxicologic significance, car exhaust, fossil fuel combustion, deforestation, municipal and household waste incineration, metal smelting and pesticide spraying contribute largely to the introduction of heavy metals into the environment. These substrates pose serious and adverse risks and threats to the very sensitive ecosystem that exists within the soil and other ecological niches. Resistant to biodegradation and biological breakdown, these metals can be checked for their presence and levels via phytoremediation strategies which have been elaborated in this chapter.

Addressing the Pressing Issues Highlighting the Need and Significance of Phytoremediation

The emergence of new and technologically advanced industrial setups such as electrical, pharmaceuticals and material science have played a largely significant role in the addition of pollutants in our biosphere. These pollutants include a variety of compounds ranging from organic to inorganic constituents that are extremely harmful to our ecosystems and human health. Organic compounds include petroleum hydrocarbons, chlorinated solvents, linear halogenated hydrocarbons, volatile organic carbons and trinitrotoluene whereas inorganic contaminants include phosphates, nitrates, metals and metalloids like mercury, zinc, lead, cadmium, arsenic, chromium, nickel and copper and radioactive or nonradioactive nuclides like strontium, uranium and caesium. These contaminants and harmful chemical constituents form a huge part of the loads of waste discharged into the rivers, lakes and streams or other water bodies and dumped in land, hence posing enormous environmental and health hazard risks. These metals form an essential part of the ecosystem but only when found in trace amounts; accumulation or aggregation of these metals in amounts hard to decompose or treat puts forth major ecological and environmental risks as these toxic synthetics and chemicals are difficult to chemically or biologically degrade or break down.

As per a current analysis, the presence of these hazardous chemicals in soil, water and air and food is a stressing global issue in terms of effective and sustainable waste management.

Microbe-associated or assisted remediation have been investigated and practised as real-time solutions for the detoxification of complex pollutants and this particular convergence of phytoremediation and bioremediation assures a sustainable approach towards the effective removal of industrial waste from land and water (Srivastava P. and Giri N., 2021).

Identification of Different Industrial Pollutants and Toxic Contaminants and Their Respective Sources

Fly Ash – Thermal Power Plants

Fly ash has been identified as a significant environmental pollutant with large quantities of it being pumped out of thermal power plants and energy generation centres. Fly ash comprises particles with size ranging from as small as 0.01 to 100 micrometres and these fine particles form a great percentage of PM (Particulate Matter). Enhanced energy requirements and non-alignment of the power generation protocols with ecological norms has created a menace with increased dependence on non-renewable resources for energy production such as coal (Srivastava P. and Giri N., 2021). The coal utilised in Indian power stations has significantly higher percentages of ash (34-35%) than other varieties of coal used worldwide. As per statistical surveys, thermal power stations release close to 100 million tons of fly ash every single year and only 10% of this total amount is remediated with the significant remainder being dumped in either landfills or settling ponds. Particulate matter smaller than 10 micrometres has been linked with fatal health risks and that smaller than 2.5 micrometres has been known to cause severe respiratory distresses. Examples of plant species that possess the ability to grow in fly ash and reduce its contamination via remediation include plants from families of *Brassicaceae*, *Chenopodiaceae*, *Fabaceae*, *Leguminosae* and *Poaceae*. Besides, *Leucanea leucocephala* and *Acacia* sp. have been identified as plants with features such as high tolerance and efficient survival in metal-polluted areas (Srivastava P. and Giri N., 2021).

Pesticides and Heavy Metals

Pesticides

Synthetic pesticides and chemical substrates form a large portion of the contaminants that are leached into the soil and pollute the land because of their constituents being heavy metals and harmful organic solvents. Extensive use of chemical synthetics and pesticides has done more damage than bring benefits as overuse of these pesticides has brought about environmental losses including metal contamination in soil, caused the disbalance of nutrients in soil, degraded soil quality and affected the overall soil health including soil pH and soil fertility. These chemicals bring short-term benefits such as increased produce quantity and quality but have long-term disadvantages such as contaminating the food and affecting the concerned food chains and webs, causing chemical resistance in pests and pathogens (due to improper use and overuse) and causing soil degradation (Srivastava P. and Giri N., 2021).

Insecticides and pesticides such as aldrin, DDT, hexachlorocyclohexane (HCH) and dieldrin are very commonly used in some Asian developing countries because of their low cost and easy availability. However, regardless of their not-so-significantly beneficial minimal advantages, they have been banned in many of the advanced and developing nations for their potent abilities of bioaccumulating in soil and causing other biological damages. Chemicals of such nature have been analysed to be resistant to degradation and leaching and hence have been singled out for global soil contamination in many of the environmental communities.

With the many long-term disadvantages of chemical herbicides and pesticides coming to light after decades of extensive usage and field application, the several toxicological aspects of the different components of these compounds have also been highlighted and a country like India which is reliant on its agriculture is also currently facing the various residual damages and losses of the chemical fertilisers such as HCH, DDT, phorate and endosulphan (Srivastava P. and Giri N., 2021).

Arsenic

Arsenic occurrence and toxicity have drawn great attention in the recent times and most specifically in the Indian scene. Since arsenic exists in the -3, 0, +3 and +5 oxidation states, it may be found in different forms and states and this also serves as the determining factor in terms of the health effects arsenic intake or toxicity may cause. Arsenic occurrence and pollution can mainly be contributed to coal burning and metal smelting. Arsenic poses major environmental risks and its accumulation through food chains and webs has far-reaching health impacts. The most common forms of arsenic include the elemental and oxide forms of the metalloid (Srivastava P. and Giri N., 2021).

Chromium

Chromium is known to exist in several varying stages from Cr^{+2} to Cr^{+6} but its most stable forms that exist in soil as well are chromium (III) and chromium (VI) which possess varying chemical properties (Srivastava P. and Giri N., 2021). Contrasting to other metals, chromium-associated hazards are reliant on its oxidation state. Most danger-posing to human health and cancer-causing in nature, hexavalent chromium has been identified as the most potent pollutant and hazardous form of chromium.

Cadmium

Cadmium, one of the most common pollutants in storm water is released into the ecosystem from several industrial sources such as electroplating industries, plastic manufacturers, alloy and battery industries and paint manufacturing units. It is known to cause several health distresses such as that of kidney. The universality of cadmium due to its presence in most of the household items in the form of a coating highlights its significance and inclusion in everyday activities thus indicating the requirement for a necessary approach to limit its usage and utilisation (Srivastava P. and Giri N., 2021).

Complex Industrial Wastes

Complex industrial waste includes all the waste produced and released from industrial forefronts including distilleries, petroleum plants, paper and pulp mills, tanneries and textile manufacturing units which is extremely significantly responsible for the contamination of the soil and water of the land. This waste is characterised as complex as its characteristics include high BOD (Biological Oxygen Demand), high COD (Chemical Oxygen Demand), phenol, sulphate, levels of heavy metals and TDS (Total Dissolved Solids). Due to its diverse composition, complex waste is difficult and challenging to dispose of safely in the environment.

Distillery Wastes

Distillery industries generate large amounts of waste consisting of heavy metals and potent contaminants. As per the statistical data, a total of 12 to 15 litres of effluent is produced per litre of alcohol produced. This effluent which is untreated is referred to as spent wash. Post-anaerobic digestion spent wash is recognised as PMDE (post-methanated distillery effluent) and is rich in BOD, COD and TDS. Besides, distilleries also release huge amounts of PMDS (post-methanated distillery sludge) which is rich in OM (Organic Matter), TDS and heavy metals.

These wastes, rich in heavy metals, upon release in water bodies are responsible for damaging the aquatic environment and disturbing the ecology of marine organisms assisted with a reduction in sunlight penetration into water bodies affecting the photosynthetic abilities of marine plants and reduction in dissolved oxygen content. Upon land disposal, these wastes inhibit vital plant processes like seed germination and cause widespread depletion of vegetation and the withdrawal of green cover assisted with a significant reduction in soil alkalinity and nutrient availability (Chandra R. et al., 2017).

Petroleum Wastes

An increase in petroleum product processing has caused enormous waste production from petroleum production units magnifying the environmental and ecological risks posed by these harmful wastes which include petroleum hydrocarbons (PHCs) which is a significant mixture of hydrocarbons, most specifically alkanes, aromatic hydrocarbons, cycloalkanes and many more complex chemicals (Chandra R. et al., 2017).

PHCs accumulation has been identified as an adverse risk to environment and humans alike. As a result of which, the remediation and removal of PHCs has gained worldwide sensitisation and attention in terms of phytoremediation. Several plant species have been singled out for their ability to remediate PHC-contaminated lands and offer for a successful reclamation. These examples include *Eichhornia crassipes* (water hyacinths), prairie grasses and other weed varieties such as *Festuca arundinacea* (tall fescue), *Panicum virgatum* L. (switchgrass), *Lolium perenne* (ryegrass), *Digitaria sanguinalis* (crabgrass) and smooth meadowgrass.

Paper and Pulp Mill Wastes

These mills are identified and highlighted as the most polluting and contaminating industries worldwide assisted with recognition as the second highest consumer of chlorine and the largest discharger of organochlorine in wastewater. The waste from these industries includes pollutants characterised with high toxicological significance such as dibenzofurans and dibenzo-p-dioxins which constitute together with other chemicals to form effluent rich in BOD, COD and TDS. This effluent is characterised with low biodegradability due to the presence of recalcitrant compounds (Chandra R. et al., 2017).

Tannery Wastes

Tannery wastes include waste from leather and leather manufacturing centres, fur production sites and retail stores which consists of a large number of by-products and solid wastes which are ultimately discharged into water bodies or dumped in landfills. Besides, these industries are held responsible for their toxic emissions into the atmosphere. The tanning industry has been reported to produce about 600,000 tons of solid waste every year which comprises all sorts of waste categories and end-products such as leather trimmings, dust, hair, fleshings and shaving dust. The Indian tanning industries contribute to about 15% of the total tannery waste produced and estimated worldwide (Chandra R. et al., 2017).

The waste originating from these sources is characteristically rich in heavy metals such as iron, chromium, cadmium, lead, zinc, manganese and copper. These metal-rich waste products pose serios threats to the environment and human health due to their ability to accumulate and resist biodegradation.

Textile Wastes

The most pressing issue concerning the developmental requirements of the textile industries around the globe is the amount of wastewater that is released as a part of the manufacturing cycle and the significant load of chemicals this wastewater is enriched with. Ranging from fibre-washing to the processes of bleaching, dyeing and final washing, textile production requires and utilises extremely large quantities of water, thus resulting in the production of increased amounts of chemical-enriched wastewater and sludge. This wastewater contains the many chemicals, colouring agents and dyes the clothing product is treated with including toxic residual leachates and intermediates. As much as 10 to 15% of the dyeing colours and bleaching agents are lost during the application process and are hence released along with the wastewater. This wastewater is characterized by high BOD, COD and TDS, indicating that major environmental and heath risks are probable if discharged sans effective and successful treatment (Chandra R. et al., 2017).

Agrochemicals

The current and modern agricultural functioning requires extensive application of synthetics and chemical fertilisers to obtain a quantity-rich and quality-fit produce. Hence, these agriculture chemicals or agro-chemicals have been recognised globally as major pollutants of waterbodies and land soils because of their characteristic richness in heavy metals including fertilisers, pesticides, fungicides and chemical plant growth regulators and pest repellents. These wastes are toxic and degradation-resistant and so, they are significantly responsible for contaminating the ecosystem and soils. Also, these chemicals, post-application, are leached into the groundwater or bottom layers of soil due to precipitation and other biological factors, causing a disturbance in the soil nutrient availability and groundwater pollution (Chandra R. et al., 2017).

Types of Phytoremediation: Associated Principles and Applications

Phytoremediation can be classified into several categories based on the contaminants, types of plants, contaminated site, technology used and the level of clean up or reclamation required. These include: phytoextraction, phytostabilization, phytovolatilization, phytofiltration, phytodegradation, rhizodegradation and phytodesalination.

Phytoremediation is a significantly diverse term involving a variety of techniques and processes for decontamination of soil, sludge, land and wastewater (Chandra R. and Kumar V., 2017).

Phytoextraction

Phytoextraction is also referred to as phytoaccumulation, phytosequestration and phytoabsorption and it indicates the use of plants for the absorption, translocation and storage of hazardous and toxic chemicals and contaminants from soil, sediments in the root and shoot tissues. Highlighted as the most efficiently recognised and practised technique of phytoremediation, the process of phytoextraction requires a long-term maintenance and regular harvesting schedule of plants assisted with a safe and secure disposal of contaminated plant materials. Since the cost involved in practising the phytoextraction technique is almost 10-fold lower than a conventional soil remediation technique, it has been widely accepted as a cost-effective technique. It occurs in the root zone of the plant and is extremely effective for the removal of metals such as arsenic, cadmium and nickel (Chandra R. and Kumar V., 2017; Srivastava P. and Giri N., 2021).

There is a set of defined characteristics that the plants ideal for phytoremediation should have. These characteristics include:

- High-level metal concentration tolerance.
- Accumulation of high levels of metals in harvestable parts.
- Wide distribution assisted with a deep root system.
- Pest and disease resistance.
- Herbivore repulsiveness.
- Easy cultivation, harvesting, processing with low requirements for cultivation.
- High growth rate and biomass production.

Biomass production and the plant's bioconcentration efficiency have been highlighted and noted as the most significant factors that play a role in determining a plant's importance as a phytoextractor. The bioconcentration efficiency of a plant is defined as its functional ability to aggregate or accumulate metals from contaminated and polluted soils and is measured using the BCF i.e., the bioconcentration factor which is the ratio of the concentration of the metal/s in the plant roots to that present in the soil or sludge. Another term, TF i.e., the translocation factor determines the efficient ability of the plant to translocate or transport the metal or metals of concern from the root to the shoot and is further elaborated as the ratio of metal accumulation or concentration in the shoot to that in the root system. BCF single-handedly plays a significantly major role in serving as a pivotal measure in terms of determining the desired potential of a specific species for phytoextraction. Any species singled out for potent phytoextraction abilities based on its BCF and TF being > than 1, can serve as a significant phytoextractor.

Here are two strategies as per which phytoextraction can be carried out: chelate-assisted phytoextraction which is also referred to as induced phytoextraction makes use of chelating agents as soil extractants and maintains the solubility of trace elements in hydroponic solutions hence, enhancing metal accumulation in plants. Several chelating agents such as EDTA (Ethylene Diamine Triacetic Acid), N-(2-hydroxyethyl)-ethylene diaminetriacetic acid (HEDTA), diethylene-tetraamine-pentaacetate acid (DTPA), ethylene glycol-O, O'-bis-(2-amino-ethyl) N,N,N',N'-tetraacetic acid (EGTA), citric acid and ethylenediamine di-o-hydroxy phenylacetic acid (EEDS) have been reported to effectively enhance the process of phytoextraction by increasing the effectiveness of metal mobilisation and metal accumulation in plants. Continuous phytoextraction makes use of hyperaccumulator plants and depends on the plant's natural functionality to carry out the accumulation, translocation and resistance of large quantities of metals throughout the entire growth cycle. Deemed as highly environment-friendly, continuous phytoextraction can be applied at two different types of sites to extract the many benefits of the process: contaminated soils found in post-industrial areas and soils located close to transportation routes (Chandra R. and Kumar V., 2017).

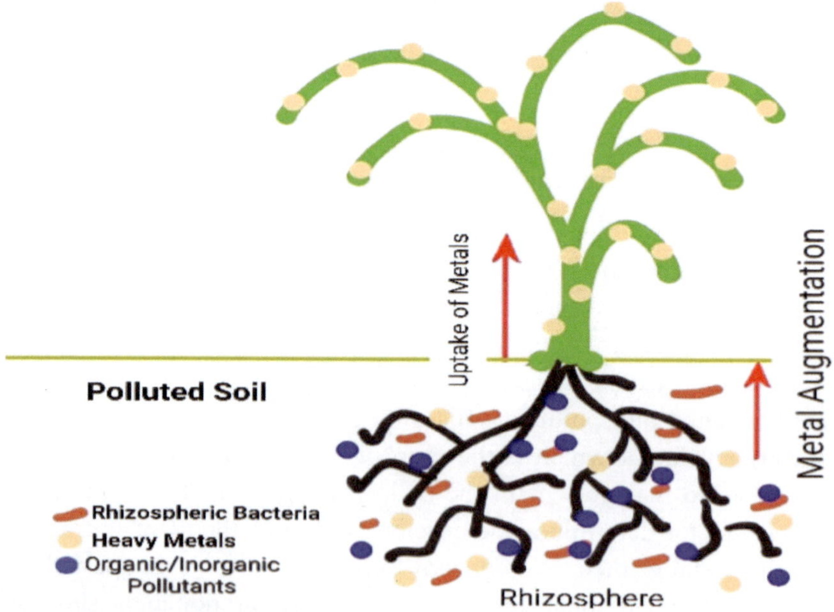

Figure 1. *Illustration of Phytoextraction*

Phytostabilization

Phytostabilization also referred to as phytoimmobilization and in place or *in situ* inactivation aims at reducing the mobility and bioavailability of pollutants and contaminants in the ecosystem. This strategy focusses on the immobilisation of the contaminants in the soil and this can be achieved by adsorption and accumulation performed by the roots, adsorption taking place on roots or precipitation within the root zone and physical stabilisation of the soil, thus, ensuring there is a reduction in the risks to environmental and human health. Several heavy metals, like lead, arsenic, cadmium, chromium, copper and zinc can be effectively immobilised by making use of the phytostabilization technique. In contrast to phytoextraction or hyperaccumulation of metals which ensures accumulation of metals into the shoot or root tissues, phytostabilization ensures the sequestration of metals and contaminants within the rhizosphere rather than in the plant tissues (Chandra R. et al., 2017; Chandra R. and Kumar V., 2017).

Certain set of characteristics a plant must have to be selected as a potent phytostabilizer include:

- Highly tolerant to high levels of contamination.
- Highly productive; rapid biomass production.
- Able to retain, reduce or precipitate the applicable pollutants in roots.

Table 1. *Commonly Reported Plants Capable of Heavy Metal Hyperaccumulation / Phytostabilization*

Plant Species	Metal/s
Isocoma veneta (Kunth) Greene	cadmium, copper, manganese, lead, zinc
Teloxys graveolens	cadmium, copper, manganese, lead, zinc
Euphorbia sp.	cadmium, copper, manganese, lead, zinc
Dalea bicolor Humb. & Bonpl. Ex Wild.	cadmium, copper, manganese, lead, zinc
Lygeum spartum L.	copper, lead, zinc
Piptatherum miliaceum L. Coss.	copper, lead, zinc
Atriplex lentiformis (Torr.) S. Wats.	arsenic, copper, manganese, lead, zinc
Atriplex canescens (Pursh) Nutt.	arsenic, mercury, manganese, lead
Bidens humilis	Silver, arsenic, cadmium, copper, lead, zinc
Baccharis neglecta Britt.	arsenic
Schinus mole L.	Cadmium, copper, manganese, lead, zinc

Examples include *Jatropha curcas* which can grow in chromium-contaminated soils and possesses the ability of accumulating chromium in roots followed by shoots. Consequently, chromium intoxication triggers oxidative stress in *J.curas* which it is easily able to tolerate via its antioxidant defence system. Other similar examples of some very commonly reported plants which possess the ability of heavy metal hyperaccumulation have been mentioned in Table 1 (Chandra R. et al., 2017).

Phytovolatilization

Phytovolatilization refers to the release or addition of contaminants which are volatile in nature to the atmosphere of the ecosystem. This takes place via plant transpiration and is an advanced form of phytotransformation which is applicable to only those metals which are highly volatile for example mercury, arsenic and selenium. A complicated process as it is, phytovolatilization is mainly and significantly driven by certain biophysical processes which facilitate the mass transport of these metals via vegetation i.e., these metals are absorbed or taken up by the root system of the plants and after a conversion of these metals into non-toxic forms they are volatilized into the atmosphere. The process of volatilization of these metals into the atmosphere can be from any of the plant parts including shoots, leaves or the roots.

Example of this technique includes *Brassica juncea* (Indian mustard) which has been observed and experimented upon as to having an extremely high rate of selenium hyperaccumulation and volatilization assisted with a fast growth ability which singles it out as a potent phytovolatilizer. Phytovolatilization is not only effective for the volatilization of metals but also significant in the removal of organic pollutants alike. Example includes *Liriodendron* sp. (poplar tree) which has been identified to volatilize amounts of trichloroethane taken up from degraded and polluted soils as high as 90% of the total.

Notably, unlike the other phytoremediation technologies, phytovolatilization ensures the complete and absolute decontamination of the concerned site assisted with the elimination of the need of harvesting or disposing of the plant species (Chandra R. et al., 2017).

Phytofiltration

Phytofiltration has been highlighted as a technology that focusses on the remediation and absorption of the contaminants from water bodies and aqueous waste streams. This technique of filtration of pollutants can be achieved via three major processes: rhizofiltration, blastofiltration or caulofiltration.

Rhizofiltration is specifically a process of the root zone as per which an interaction between terrestrial and aquatic plants and the wastewater pollutants takes place. Significantly efficient when low quantities of pollutants and large amounts of water are involved, rhizofiltration immobilizes the contaminants, thus preventing their seepage into the groundwater. Characteristics that are necessary for a plant to be selected as an ideal rhizofilter include:

- Rapid root growth.
- Toxic metal removal over prolonged time periods.
- The principles which determine the efficiency of phytofiltration include precipitation, adsorption, bioaccumulation and rhizodegradation (Chandra R. et al., 2017).

Caulofiltration technique shifts the focus towards the *caulis* i.e., the shoot for the purpose of removal of contaminants from the wastewater. In an analytical study observation, it was found that upon the adsorption or absorption of the metals, the mobility of these contaminants was restricted in the underground water.

Example includes *Berkheya coddii* which has been highlighted as a nickel-accumulating plant species, thus qualifying for the phytoextraction and phytofiltration of metals from wastewater.

Blastofiltration technique highlights the role of plant seedlings in the process of removal of contaminants from wastewater and aqueous solutions. As per this technique, the ability of the seedlings to extract huge amounts of heavy metals and the post-germination rapid increase in the volume-to surface area ratio are employed to make this mechanism suitable for the remediation of wastewater polluted with industrial remnants and pollutants.

Examples of the seedlings capable of performing this technique include sunflower, pea, castor oil plants, corn, Chinese sorghum and Indian mustard (Chandra R. et al., 2017).

Phytodegradation

Alternatively referred to as phytotransformation, phytodegradation ensures the metabolic breakdown and transformation of the pollutants and contaminants absorbed by plants. This is induced by the effect of enzymes produced, secreted and released by the plant itself to ensure the chemical alteration of the metals close to the plant. Pollutants and specifically organic contaminants are worked upon by these secreted enzymes under the metabolic processes which are referred to as *ex planta* and converted into simpler absorbable forms which eventually aid in the plant growth and mechanism. The factors which significantly determine the metal uptake into the plant system include the efficiency of the uptake, the rate of transpiration and the chemical or metal concentration in the soil and water.

Examples include certain specific enzymes which possess the ability to chemically transform and breakdown waste categories like chlorinated solvents (trichloroethane), herbicides and ammunition wastes, petrochemical contaminants, fuel spills, agricultural pollutants and landfill leachates. The direct advantage of this technique is that the derived or newly formed compound is nontoxic or not as toxic as the parent substrate.

Rhizodegradation

Rhizodegradation is a technique which facilitates the substrates supplied by plants to initiate the growth of microbial populations in the rhizosphere to provide for the breakdown of soil organic contaminants. Microorganisms are capable of thriving in the rhizosphere which serves as a suitable and favourable niche for such growth. These microbes convert and transform harmful pollutants into non-hazardous products which cause minimal or no harm at all. Slower than the process of phytodegradation, rhizodegradation functions on the ability of the microorganisms to increase in number and enhance their activity in the presence of certain nutrients which are secreted and released in the form of exudates from the plant roots. The characteristics of rhizodegradation that make it an efficient technique and approach towards successful phytoremediation include elimination of contaminants *in situ*, absolute mineralisation of organic metals and pollutants, enhanced plant growth via nutrient fixation, mobilization, protection from plant pathogens, pollutant degradation and decreasing the plant stress hormone levels (Chandra R. and Kumar V., 2017; Srivastava P. and Giri N., 2021).

Phytodesalination

The technique of phytodesalination makes use of halophytic plants for the purpose of remediation and removal of contaminating salts from saline lands and areas. With adjustments and adaptations including osmolarity maintenance and compartmentalization of ions in the vacuoles, halophytic plants have been investigated and highlighted to function significantly efficiently in saline environments including sea water, salt desert or salt marsh. As per the most recent estimates and publishing, two examples have been specified as extremely effective and successful halophytic remediators and these include *Suaeda maritime* and *Sesuvium portulacastrum* which have been identified to remove salt quantities as high as 506 and 476 kg from 1 hectare of soil within a duration of 4 months (Srivastava P. and Giri N., 2021).

Plant Categories with Growth Capability in Metal-Polluted and Degraded Areas

Plant species have been identified to work as per three basic strategies of survival which facilitate their growth in metal-polluted sites. These include:

Excluders

Excluders are referred to as plants which possess the ability of facilitating the limitation of translocation of heavy metals within them and the maintenance of low levels in shoots over an extended range of soil and wastewater contaminant levels. This aids in the protection of the tissues of leaf, most specifically the ones comprising the essential photosynthetic cells. Nevertheless, these plants can contain large amounts of metals in their roots.

Indicators

In contrast to the excluders, indicators are capable of accumulating heavy metals in the shoot tissues. This is also indicative of the levels of metals in the soil or wastewater. But, upon continuous and prolonged periods of metal uptake, these plant species die. Since they provide a direct or indirect information of the effects of pollutants on the environment and even absorb pollutants similar to accumulators, indicator species are of enhanced ecological significance. The ratio of the metal in plants to that of soil is 1.

Accumulators

These species work in the direction of accumulating metals in the shoot tissues in levels exceeding than that of the soil. Upon continuous uptake and accumulation, these plants tend to die. The ratio of the metal in plants to that of soil is > 1.

Phytoremediation Mechanisms Devised by Plants

Heavy Metal Tolerance

Heavy metal tolerance in plants is highlighted as the plant's ability to survive, thrive and perform its functional mechanisms in a soil which is characteristically toxic to other plant species. Physical barriers act as the first line of defence against heavy metals. Essential amounts of metals are known to be found between the cell wall and the cell membrane. Besides, pectin plays a major role in the process of response to metal stress. Divalent and trivalent heavy metal ions attach with the carboxyl groups of pectin and this causes the restriction of the movement of metals to the plasma membrane.

Further, the plasma membrane restricts the entry of these heavy metal ions into the cytoplasm ensuring no contamination takes place. This is done by certain mechanisms which prevent the binding of the metal ions to the plasma membrane and these include alteration of the cell wall binding capacity to the ions of the metals, modification of the ion channels present on the membrane and modification of the efflux pumps.

Another mechanism includes the synthesis of cellular biomolecules which will counter the effect of the metal toxicity. These biomolecules include nicotianamine, spermine, mugineic acids, glutathione, metallothioneins, putrescine, organic acids, flavonoids, phenolic compounds, phytochelatins, histidine, hormones, heat shock proteins, salicylic acid, ethylene and jasmonic acid.

Accumulation and Detoxification

Plants possess the natural ability of accumulating heavy metals which include magnesium, iron, zinc, manganese, copper, molybdenum and nickel and utilising these metals for their own growth and development. However, some plants also tend to accumulate metals like cadmium, chromium, lead, cobalt, silver, selenium and mercury even though these metals have no essential functions in plant metabolism (Ugya Yunusa A., 2016). A rare class of plants known as hyperaccumulators accumulate trace elements as well as other metals due to their high resistance to metal toxicity and the capability of accumulating metals in the leaves. These types of plants have a significant role in the process of phytoremediation due to these benefits. They do this by accumulating large percentages of metals in their aerial parts without showing any signs of intoxication. Over time, plants have been known to adapt certain strategies to exhibit flexibility in even adverse conditions (Lin Lang-Dong et al., 2021). These strategies include modification of the gene expression, alteration of the membrane transporters that help in the processes of vital importance such as efflux, translocation, sequestration and uptake. Hyperaccumulator plants portray advanced features of absorption of large amounts of metals and being able to translocate and hypertolerate them. Processes of chelation and sequestration help in the elimination of hazardous toxic ions from sensitive sites. The metals which are hyperaccumulated are ultimately bound to ligands such as histidine, citrate, malate, phytochelatins and glutathione which facilitate the detoxification and sequestration processes in plants.

Conclusion

Phytoremediation has gained enhanced acceptance as a plant-based and sustainable technology working effectively in the direction of sustainable waste management of toxic and difficult to degrade or eliminate industrial pollutants and contaminants. Several industries and organisations have come forth in an effort to restore the lost ecological significance of areas subjected to metal toxication and degradation via remediation and reclamation efforts. Three nations namely – Kenya, the UK and the US have highlighted the concept of green belts which refers to "a strip of trees of such species and such geometry that when planted around a source would considerably attenuate the pollution in air by intercepting and accumulating the toxic elements in sustainable manner. Selection of such lands and even selection of the suitable plant species plays an essential role in establishing the concept of green belts. This methodology can aim at effective removal of contaminants from the environment and reduce pollution drastically. Planting vegetation in the areas of landfills has been idealised as the most viable technique in terms of remediation which will help in several ways such as providing stability against wind and water erosion, decontamination of the concerned land area and the aesthetic enhancement of the area. Environmentally friendly and cost-saving techniques as it is, phytoremediation, in the current scenario is attracting global attraction and acceptance and will lead to a broader sense of generalised awareness in terms of sustainable waste management approaches

The wide spectrum of possibilities achieved with phytoremediation techniques in laboratory-specific conditions are promising but vary greatly from the field application. Scientists and ecologists have believed alike that the field application of these methodologies requires more understanding of certain basic concepts such as the levels of metal amounts that can be absorbed and accumulated by the concerned plant species and the limited availability of the concerned pollutants and contaminants as the metals are present in insoluble forms in the soil, making them less available unlike that of a laboratory setting. The development and awareness about genetically engineered plant species can bring to light more real-time solutions for these shortcomings. This has invoked manifold enhancement in the field of research in an attempt to target, analyse and comprehend the very multifaceted phenomenon of metal absorption, accumulation, volatilization, mobilisation, remediation, and sequestration making this green technology more collaborative and interdisciplinary. Additional studies will invite enhanced and rigorous research into tailoring plant species for remediation purposes and more perspectives of the microbe-assisted outlook of phytoremediation.

Disclaimer

None

References

Ahmad P. *Plant Metal Interaction* (2015) ISBN: 9780128031582.

Chandra R. and Kumar V. Phytoremediation: A Green Sustainable Technology for Industrial Waste Management. *Phytoremediation of Environmental Pollutants* (2017) ISBN: 9781351665636.

Chandra R., Kumar V., Tripathi Sonam, Sharma Pooja. Phytoremediation of Industrial Pollutants and Life Cycle Assessment. *Phytoremediation of Environmental Pollutants* (2017) ISBN: 9781351665636.

Lin Lang-Dong, Ho Jian-Ren, Yang Bing-Yuan, Ko Chun-Han and Chang Fang-Chih. Life Cycle Assessment of Heavy Metal Contaminated Sites: Phytoremediation and Soil Excavation. *International Journal of Phytoremediation* (2021) 24(4):334-341.

Pandey V. C. and Bauddh K. *Phytomanagement of Polluted Sites* (2018) ISBN: 9780128139127.

Roy M. and Pandey V. C. Role of Microbes in Grass-Based Phytoremediation. *Phytoremediation Potential of Perennial Grasses* (2020) ISBN: 9780128177327.

Srivastava P. and Giri N. Role of Plants in Phytoremediation of Industrial Waste. *Bioprospecting of Plant Biodiversity for Industrial Molecules* (2021) ISBN: 9781119717218.

Ugya Yunusa A. *The Role of Phytoremediation in Remediation of Industrial Waste*. (2016) ISBN: 9783668351233.

Chapter 6

Assessing Indian Legal Framework to Address Sustainable Waste Management

Ajai Prakash[1], Ajitesh Tiwari[1] and Jyoti Prakash[2*]

[1] Department of Business Administration, University of Lucknow, Lucknow, Uttar Pradesh, India

[2] Amity Institute of Biotechnology, Amity University Uttar Pradesh, Lucknow, Uttar Pradesh, India

Abstract

The burgeoning population of India requires production and consumption to be expanded at an unprecedented pace in order to meet the needs of population, which has been the case in the recent decades. This poses an enormous challenge of minimizing wastage of resources and handling the waste in least harmful manner. While individual and community led initiatives may help to mitigate the burden of waste management, the primary responsibilities lie with the public authorities. These authorities require an apt regulatory mechanism to deal with the challenges of handling diverse type of wastages. This piece of research reviews a set of legal provisions beginning from general provisions initiated with the Wildlife Protection Act-1972 to highly evolved and specified regulations such as handling of Construction and Demolition Waste Rules-2016, Plastic Waste Management Rules-2016, Hazardous Waste Management Rules-2016, Medical waste management rules, Industrial waste Management rules, electronic waste management rules etc. In addition, it also analyses important legal principles evolvedto regulatespecific identities such the polluter pay's principles for those who pollute individually, and Extended Producer Responsibility (EPR) 2022 for producers.

Keywords: Waste Management, Indian legal framework, rules.

* Corresponding Author's Email: jprakash@lko.amity.edu

Introduction

India, is a developing country strives to secure better living standards to roughly one sixth of the global population, evolving standards of environmental protection is essential to ensure sustainability of development efforts while ensuring congenial conditions for survival of natural and human habitat. The Indian policymakers have been aware of this imperative to design and implement a framework to guide all stakeholders in this direction. Innovative techniques such as phytoremediation and biological filters are being used to treat heavy metal contamination to treat hazardous waste (Prakash 2013).

The Indian Constitution and Environment

Environment as a subject has been drafted in the Indian constitution. Outlines of the issues covered under the Indian constitution are discussed as follows.

Article 48A in the Constitution of India

The directive principles of state policy under article -48A ordains the state as "The State shall endeavor to protect and improve the environment and to safeguard the forests and wild life of the country". This directive was inserted by the 42nd amendment in the constitution of India in 1976 and the same continues.

Further, outlining the fundamental duties of the citizens, the Article 51-A (g), states that "It shall be duty of every citizen of India to protect and improve the natural environment including forests, lakes, rivers and wild life and to have compassion for living creatures." This outlines that both, the citizens and the state has been vested with the responsibility of taking care of the environment under the constitution.

The various facets of the subject of environment have been spread under different authoritative jurisdictions in the Indian Constitution. Air has been put up under the Union list, Water under the state list and forest in concurrent lists. For example, the concurrent list of Entry 17 states that, water is the subject of the States. Entry 17 under List II of Seventh Schedule provides that "Water, that is to say, water supplies, irrigation and canals, drainage and embankments, water storage and water subject to the provisions of Entry 56 of List I". The residuary powers i.e., those issues which have not been covered under the state or the concurrent list are taken care of by Article 248 of the Constitution of India. The residuary powers of legislation which states that the "(1) Parliament has exclusive power to make any law with respect to any matter not enumerated in the Concurrent List or State List. (2) Such power shall include the power of making any law imposing a tax not mentioned in either of those Lists." This also ensures that taxes such as carbon tax can be imposed by the government to protect environment.

One of the major changes that was instituted in the understanding and application of environment sensitivity was brought about in 2014 with the name of the Ministry of Environment and Forest changed to The Ministry of Environment Forest and Climate Change. This implied more serious efforts towards abatement of climate change.

International Step-ups and Their Enactment in India

The Club of Rome, which was an informal group involving stakeholders from various aspect affecting the planet earth, was an initiator of various issues on environment in 1968. The meet was followed by a report titled "Limits to Growth". This was published in 1972 when the Stockholm Conference on Human Environment was held in June 1972. The then Indian Prime Minister of India, Indira Gandhi participated in the same. She outlined poverty as the main cause of Environmental Degradation.

In pursuance of the commitments made in Stockholm conference, the Indian Parliament enacted the Wild Life (Protection) Act in 1972. It delineated the boundaries of the protected areas such as National Parks and Wildlife Sanctuaries while prohibiting activities that may interfere with the serene milieu of these landscapes or generate waste which could have been disturbing for the ecology of these areas.

The Water (Prevention and Control of Pollution) Act, 1974 and The Air (Prevention and Control of Pollution) Act, 1981 created regulatory structures such as Central Pollution Control Board (CPCB) and the State Pollution Control Boards (SPCBs) in different states to evolve procedures for preventing pollution and generation of wastages which create problems of pollution. The concept of sustainability has witnessed a widened scope of sustainability under various subsectors and there are numerous principles evolving in the banking and finance domain (Waseem 2020).

The Structure of Environment Protection in India

The Bhopal gas tragedy of 1984 created a national issue and a serious consideration. The India legal framework gave a deep thought on environment protection and laws and regulations were initiated and created to channelize wastes. Wastes were segmented and weights assigned in order of requirements under the laws.

The then government under the prime minister ship of Rajiv Gandhicame up with necessary enactments in the Indian Legislation. Environment Protection Act 1986 (EPA1986) came up, which was applicable to the whole of India. There are

26 sections under the EPA1986, divided into 4 chapters and has its genesis in Indian Constitution's Article 48(A) and Article 51(A)g. Enacted with the powers vested under Article 253 of the Indian Constitution. It states that *"it has the sole aim of ensuring the protection of the environment, the prevention and reduction of environmental pollution and provides the authority to take strict action against perpetrators.*

This Act takes precedence over other Acts which implies that in addition to other acts this act too has its periphery of application of laws and will be given the highest priority. Some of the rules formed for waste management under the act are discussed hereunder.

Salient features of the EPA1986

- The Act covers all forms of pollution; air, water, soil and noise.
- It provides the safe standards for the presence of various pollutants in the environment.
- It prohibits the use of hazardous material unless prior permission is taken from the Central Government.
- It allows the central government to assign authorities in various jurisdictions to carry out the laws of this Act.

Important Chapters and Sections of the Environment Protection Act, 1986

Chapter II EPA- General Powers of the Central Government, section 3 specifies the following points, among others:

- The central government has the authority to lay down the standards for different pollutants, and also the safe limit of emissions or discharge of wastes from industries.
- It can restrict the areas in which industries can operate.
- It has the power to lay down the safety measures and procedures for the prevention of accidents, and specify the protocol and remedies if such an accident occurs.
  - The central government has the responsibility to carry out and fund research on environmental pollution.
  - It needs to establish environmental laboratories.

It is responsible for the collection and dissemination of information related to environmental pollution.

Plastic Waste Management Rules 2016

The recent estimates suggest that plastic waste generation in the country is more than three million tones and the recycling is not even half of this quantity (Down to Earth 2021). Realizing the alarming situation of the mammoth quantity of waste generated and the consequent toxicity, Ministry of Environment, Forest and Climate Change (MOEFCC) first introduced the Plastic Waste Management Rules, to provides a planned structure for sustainable plastic use, it for the first time provided for the application of the concept of extended producer responsibility and changes in product composition to eliminate the single use plastic, from the country, which is considered to be the most polluting agent in terrestrial and oceanic ecosystems. The Plastic Waste Management Rules subsequently have been amended in 2018 with recent amendments introduced in 2021 which have become effective from the 2022.

The new rules include inert alia:

Categorization of Plastics according to usability and material composition i.e.

- Category1: Hard plastic packaging is covered under this category. This is suitable for reuse and provided for such usage in the rules.
- Category2: This category will include multilayer or single layer plastic which might contain plastics of varying chemical composition. The other plastics under this category are frequent use items like plastic sheets, sachets, pouches commonly used for commercial packaging of consumer goods.
- Category3: Packaging materials which combine plastic and non-plastic packaging are to be included in this category which essentially means that it is a multilayer plastic which consists of at least one non plastic material layer.
- Category4: This category would include packaging material made up of compostable plastics.
    - Light weight plastic wastes have been havoc to the environment. With effect from 30th September, 2021 the legislation for preventing littering of plastics has been enacted. The legislation looks at the thickness of plastic and defines the micron size of the plastic to prevent usage. a large use of carry bags with up to seventy-five microns has been increased from fifty microns. Further with effect from the 31st December, 2022. the thickness for usage is being increased to one hundred and twenty microns Such a thick micron plastic can be reused for recycling in a better manner due to increase in thickness (Vikaspedia 2022).

- With the intent of phasing out the single use plastic from the entire supply chain, the rules also ban on production, importing, warehousing and marketing, sale and end use of twenty-two types of single use plastics from the effect of 1st July, 2022. These inter alia include ear buds with plastic sticks, plastic flags, plates, cups, glasses, candy sticks, ice-cream sticks, plastic sticks for balloons, polystyrene (Thermocol) for decoration, cutlery such as forks, spoons, knives, straw, trays, wrapping or packing films around sweet boxes, invitation cards, and cigarette packets, plastic or PVC banners less than 100 micron, stirrers etc.
- Strengthening, the Extended Producer Responsibility framework the, rules provide for the sale and purchase of Extended Producer Responsibility certificates which are in surplus with an entity, it thus makes the provision for development of a market mechanism to enforce the minimum recycling levels through EPR.
- In order to provide for deterrence to for negligence of these rules, the rules provide for environmental compensation based on the polluter pays principle when a manufacturer, importer or brand owner is unable meet the Extended Producer Responsibility targets to control plastic pollution from the operations of her concern.
- For effective monitoring and compliance of the provisions of the rules, the rules provide for the establishment of a digitized platform where the responsible stakeholders in the plastic packaging and delivery supply chain would submit their annual returns and could trace the important orders and instructions with regard to the rules issued.

Construction and Demolition Waste Management Rules-2016

Another area of concern in a developing country with expanding population is construction. India too is on an expansion binge and expanding construction activity is witnessed throughout. Construction has been attributed with the status of highest pollution activity status specially during the measuring of air pollution. Construction activity in different parts of the country has been stopped during winters in different instances due to poor Air Quality Index readings.

According to the Building Material Promotion Council (BMPTC), "India generates an estimated 150 million tonnes of construction and demolition (C&D) waste every year". Official recycling capacity in India is 6,500 tonnes per day (TPD) which as a percentage is about 1 per cent. Many industry estimates put the figure at three-five times more than the official estimate (CSE 2020).

The Ministry of Environment Forest and Climate Change notified the Construction and Demolition Waste Management Rules on 29 March 2016 and the following salient features of the same are indicated below.

- The waste generated has been classified in to four different types i.e., concrete, soil, steel, wood and plastic, bricks and mortar. It mandates for segregation of these waste types for alternative reutilization purposes.
- Categorization of Waste Generators: The rules identify the waste generators such as real estate developers and service providers and contractors who generate 20 tons in a day or 300 tons per project in a month as large waste generator. The liability of preparing and submitting a waste management plan and getting it approved from the competent authority is on the waste generator.
- Relevant charges would be applicable for collection, transportation, storage and disposal to create a self-sustaining system of construction and demolition of waste management ecosystem.
- State Pollution Control Board (SPCB) and Central Pollution Control Board (CPCB) are required to grant authorization to construct and demolish waste processing facilities and provide additional guidelines for establishment of waste processing facilities.
- Local authorities like municipalities are given significant implementation responsibilities for organizing a network of collection, segregation, transportation and recycling facilities.

Biomedical Waste Management Rules-2016

As weare party to one of the biggest human masses across the globe, we are also party to contributing biomedical wastes. The biomedical waste rules have gradually evolved over a period of last two decades i.e., the first biomedical rules were enunciated in 1998 with the latest and most comprehensive one promulgated in 2016. The covid-19 has highlighted an even more serious and urgent requirement for safely and sustainably managing biomedical waste to save the communities and individuals from a range of contagious diseases and in checking the spread of the infections that may spread from such wastes.

The term biomedical waste encompasses a range of waste material generated during medical treatment, precautionary immunization, and surgical procedures as well as during diagnostic check-ups. It includes human and animal bodily waste like blood, damaged organ tissues, bodily fluids etc as well as medical devices like syringes, surgical instruments, containers, viols etc. Approximately 619 tonnes of biomedical waste per day were generated by 3,22,425 healthcare facilities during the fiscal year 2018–2019 (Saxena, Pradhan and Kumar 2022). With covid-19 pandemic vastly aggravating the waste generation across all states of India (Dehal, Vaidya and Kumar 2022).

Application and enforcement of biomedical waste management rules are essential to meet the enormous waste generated in India's medical facilities. The Biomedical waste management rules -2016 the rules cover a wide range of health care facilities including health camps, inter alia includes the following major provisions in this regard.

- Categorization of Waste: the rules provide for color code-based typification of waste to easily segregate different types of waste and define treatment and handling procedure for each one separately.

  The four categories are:

  - Yellow: waste materials covered under this code are bodily fluids of humans and animals, human anatomical waste, soiled waste, waste chemicals resulting from treatment procedures, expired medicines etc.
  - Red: plastic wastes including catheters, bags, pipes, containers, bottles etc and recyclable contaminated wastes are included in this category.
  - Blue: body implant materials, cytotoxic waste, medicine vials, glasses containing contagions are classified in this category.
  - White: Sharp metals, needles, blades, syringes with fixed needles and other materials are included in this category.
- Process Guidelines: The rules provide for necessary treatment of potential infectious biomedical waste in the premise of the health facility itself in the form of disinfection or sterilization.
- Facilities for biomedical waste treatment: The rules mandate creation of structures for specialized treatment of biomedical waste on a large scale. The rules of biomedical waste management direct that, the health facilities should not establish an additional treatment facility if any biomedical waste treatment facility exists within a radius of seventy kilometers of their location. It also mandates that the management of the common bio medical waste treatment facility would be responsible for the timely collection of such waste from health facilities located in its service area.
- It also seeks to integrate the safe and proper on-site handling of biomedical waste through training of personnel in health facility for safe and sustainable waste treatment.
- Adequate and robust guidelines have been provided for incinerators to operate them in an environmentally sound manner.

Extended Producer Responsibility (EPR)

It is very important to explicitly state the responsibility of the producers and in the same a set of rules have been instituted under the Extended Producer Responsibility (EPR) which "extend the responsibility of Producers beyond consumer stage for collection, storage, transportation and environmentally sound dismantling and recycling of E-Waste, and creating awareness among consumers through the instrument of EPR Authorization (EPRA)" (CPCB 2022)

Electronic Waste Management Rules-2016

India is also party to adding huge amounts of e-wastes. It is the third largest e-waste generator (Stista 2022) across the globe in 2019. A total of 10,14,961 tones e-waste was generated in 2019-20 as stated by CPCB. E-Waste management is based on the principle of Extended Producer Responsibility (EPR). Under EPR, once the electronic products have finished their usage and are near to their disposal it becomes the responsibility of the producers of electrical and electronic equipment (EEE-21 numbers in two categories) listed in the Schedule-I of rules to manage their products for disposal.

After collection producers have to get the E-waste dismantled and recycled in an environmentally-sound manner.

The rules provide for segregation of electronic waste into two categories which include the waste generated upon completion of the life cycle of electrical equipment as well as from electronic equipment's which are being used in information technology and other communication services. These rules include 21 items which contain metallic and chemical residues which need to be recycled in a manner which ensures optimum recycling, alternative uses and safe disposal of the hazardous pollutants which may contaminate the soil, water and air. Therefore, the entire notion of managing waste sustainably is the mainstay of the policymakers and the effort in this direction are in the nature of introducing sustainable e-waste management while ensuring that needs of the industry needs are catered.

The rules inter alia include the following provisions in this regard. The rules recognize the primary responsibility of the value chain partners in the electronic supply chain and therefore fix the responsibility of producers, wholesalers, distributors and the resellers to collect back the electronic products at the end of their life cycles and sustainably recycle them.

It provides four different schedules.

As per schedule-I, the applicability of the rules, has been extended not only to finished products being used but also as components, consumables, spares etc.

used in the electronic value chain. It also includes electrical products that contain harmful chemicals like mercury as in the case of compact fluorescent lamps.

The principle of extended producer responsibility is the single most distinguishing feature of these rules. Under these rules the extended producer responsibility has been defined as the responsibility of the producer to collect and recycle the product at the end of its life-cycle. Therefore, for collection an elaborate mechanism has been introduced in the rules. For example, the rules provide for collection centers, collection points and the mechanism such as take back systems at the end of the life cycle of the product. Therefore, it effectively seeks to implement the extended producer responsibility principle in case of e-waste management.

To drive consumers towards recycling of equipment's, it also provides for innovative options and incentivization of consumers through deposit refunds schemes, e waste exchanges, retailer and establishment of stores so that consumers are also incentivized to partake in recycling of the e-waste.

The rules also recognize that cent percent collection cannot be insured at once and therefore as per schedule-III, the rules provide for phase wise capacity augmentation for e -waste gathering and reprocessing, for instance the rules provide that under extended producer responsibility the producers have to enlarge the capacity for recycling and collection up to 70% by 7th year of the implementation of these rules.

A deposit refund scheme mandated in the rules to incentivize both consumers and producers and effectuate the recycling of e -waste under the scheme. The producer can collect an additional amount as a deposit at the time of sale of the equipment and return it to the consumer with interest when the consumer returns the electronic product at the closing stages of its life cycle.

The Central Pollution Control Board and the State Pollution Control Board are vested with the responsibility of overseeing the national and state level implementation of these rules and issue guidelines in pursuance of the rules to effectuate the rules in letter and spirit.

It also fixes the responsibilities of urban local bodies to collect and channelize the products which do not have a producer's name or where in the producer cannot be located and the local bodies may set up different dismantling or recycling facilities as their own ventures.

Hazardous Waste Management Rules 2016

Hazardous waste refers to a waste which maybe containing potentially polluting and damaging elements which may cause harm to human health and environment example, the physical, chemical, biological, toxic, flammable explosive, corrosives etc. Such elements which originate or emerge in the manufacturing process, the transportation of such substance or for end use of materials, are included in hazardous waste.

Much of the waste emanating from various industrial activities and end of the life cycle of different products containing chemicals and gases etc is treated unscientifically and disposed unscientifically. Such disposal has potential of contamination of water bodies leaching into the soils and entering into human and animal food chains at the end. These may cause different problems to human health including serious neurological diseases and genetic defects.

The Hazardous Waste Management Rules 2016 provide for scientific disposal of hazardous waste products. The Hazardous Waste Management Rules are in consonance with Basel Convention Rules on Trans Boundary Movement of Hazardous Waste.

The following measures are included in Hazardous Waste Management Rules.

The rules provide for a single window clearance system for setting up a hazardous waste management facility.

It also provides for rigorous documentation procedure for import export of waste and regulate the waste mandated for import and export through continuous revision of the list of Waste materials allowed under export and import rules.

Certain less harmful products such as metal scrap, paper waste and various categories of electronic equipment's which are meant for reuse purposes have been exempted from the need for obtaining the permission under the Hazardous Waste Management Rules 2016.

It lays down the standard operating procedures for infrastructure to manage waste processing industry and sustainably provide for safeguard human health and environment.

Under these rules, the State Governments of India have been provided various powers to implement measures which contain the spread of hazardous waste and ensure safe and environmentally sustainable collection and recycling of hazardous waste.

The state pollution control boards under these rules are mandated to furnish an annual inventory of the waste that has been generated and various stages of recycling, recovery and reprocessing that it has been gone through and submit it to the Central Pollution Control Board every year.

The state governments are also required to register all industrial spaces and facilities that generate hazardous waste and also registered the workers that are involved in the handling of the hazardous waste and those workers who which are involved in the recycling activities or reutilization activities of the hazardous waste and to undertake training and sensitization programme of these workers in accordance with the rules.

Therefore, the rules undertake a holistic view of the entire value chain of the hazardous waste and aim at preventing, minimizing, reusing and recycling as well as safe disposal of hazardous waste.

National Green Tribunal as the Apex Adjudicatory body

While these rules are being enforced by the respective central and state bodies that is state pollution Control Board in central pollution Control Board there requires to be an adjudicatory body which can handle disputes between the enforcement authorities and the producers consumers and the public at large therefore there arose a need for effective adjudicating authority backed by the force of law to adjudicate the matters which arise in these domain therefore the Government of India under the larger ambit of the climate legal responsibility and measures to establish the national green tribunal to handle these issues separately which pertain to the wider issues of environment management.

The National Green Tribunal (NGT) was set up under the national green tribunal Act 2010 as an adjudication body for any kind of disputes pertaining to compliance of the laws or other general environmental violations. As the Indian Courts are not specialized in the area of environment hence it was thought fit to initiate a body in the form of NGT. The purpose of the act mentions that the tribunal has been setup for effectual and swift clearance of cases pertaining to Environmental safety and security of forest and other natural capital including the enforcement of any legal right relating to environment and giving relief and compensation to the damages for damages to the persons or property affected adversely. The green tribunal operates under the principle of natural justice and is not bound by the procedure of Civil Code.

The national green tribunal is vested with wide powers keeping in mind the environmental justice that has to be delivered therefore the NGT has all the powers of a civil court and the disputes from it lie into High Court and not in the lower courts.

Various judgments of the tribunal have shown that the issue of enforcement and adjudication of environmental laws is being taken seriously by Indian judiciary and the administrative system and all-inclusive lawmaking agenda has been arranged by the Indian authorities to tackle the issues of environmental pollution.

Conclusion

The additional waste in India is increasing environmental pollution. While the innovation and entrepreneurship ongoing drives in the last few decades in the country have spurred the innovative spirit of our entrepreneurs. In case of India there is requirement of policy intervention to bring together stakeholders and establish the connection between waste generating and processing sectors (Tripathi 2022). This has led to the introduction of a varied range of products into markets, as a result, there is a serious need to effectively manage the of the end-of-life cycle of these products and therefore the Indian legal framework in the area of waste management has to evolve keeping in view the evolution of the nature of waste being generated in our country.

Disclaimer

None

References

CPCB. (2022, August 21). https://cpcb.nic.in/openpdffile.php?id=UmVwb3J0RmlsZXMvMTQwM18xNjU1MzU0NzkxX211ZGlhcGhvdG8xNjQ3MS5wZGY=.

CSE. (2020 August 25,) *Centre for Science and Environment.*. https://www.cseindia.org/india-manages-to-recover-and-recycle-only-about-1-per-cent-of-its-construction-and-demolition-10326#:~:text=New%20Delhi%2C%20August%2025%2C%20202-0,just%20about%201%20per%20cent. (accessed July 30, 2022).

Dehal, A., Vaidya, A. N., & Kumar, A. R. (2022). Biomedical waste generation and management during COVID-19 pandemic in India: challenges and possible management strategies. *Environmental Science and Pollution Research*, *29*(10), 14830-14845.

Down to Earth. (2021, December 21) *Down to Earth.* https://www.downtoearth.org.in/news/governance/as-told-to-parliament-december-20-2021-3-4-million-tonnes-of-plastic-waste-generated-in-2019-20-80778 (accessed July 28, 2022).

Prakash, J and Awasthi, G (2013). Accumulation of Heavy Metals in Different Water Bodies by Biological Source Algae. *Adv Life Sci*, *2*(1), 10-14.

Saxena, P., Pradhan, I. P., & Kumar, D. (2022). Redefining bio medical waste management during COVID-19 in India: A way forward. *Materials Today: Proceedings*, *60*, 849-858.

Stista. (2022, August 21). https://www.statista.com/statistics/499952/ewaste-generation-worldwide-by-major-country/.

Hussain, C. M., Singh, S., & Goswami, L. (2021). *Emerging trends to approaching zero waste*. Elsevier.

Kumar, K., Prakash, A., & Khan, W. (2020). Integrating the notion of sustainable development in banking: Analysing historical and conceptual framework. *Indian Journal of Economics and Development*, 16(3), 449-458.

Chapter 7

Current Scope of Bioinformatics in Environmental Sustainability

Shruti Singh, Akriti Bagchi and Ruchi Yadav*

Amity Institute of Biotechnology, Amity University Uttar Pradesh, Lucknow Campus, Lucknow, UP, India

Abstract

Bioinformatics is the field of computational science that amalgamates different disciplines like biology, chemistry, physics, mathematics, computer science, information technology etc. to solve various biological problem. Since decades bioinformatic has been used in drug discovery, drug target identification, understanding genes and function, system biology etc. Currently the power of bioinformatics and computational biology has been used in environmental sciences and to uncover environmental biology. Bioinformatics needs the study of microbial genomics, proteomics, system biology, computational biology, phylogenetic trees, data mining and operations of major in – silico tools for determining and analyzing the structural properties and biodegradative pathway mechanism of these harmful xenobiotic compounds. Ecoinformatics is an amalgamation of the elucidated principles of bioinformatics and the data processing of the ecological information presented by nature. This chapter discuss about the role of bioinformatics in environmental sustainability and sciences. By analysing the nucleotide sequence of genes, genomes, metagenomes, transcripts, transcriptomes, and meta transcriptomes, environmental genomics combines the diverse domains of knowledge about ancient and modern species and ecosystems. Environmental genomics offers a wealth of data on the taxonomy and diversity of living and extinct organisms, their phylogeny and evolutionary history, their capacity for adaptation and acclimatisation, their biology, their functional traits, and their interactions with the environment in both its biotic and abiotic dimensions.

Keywords: Bioinformatics, Environmental Sustainability, Genomics, Proteomics, Bioremediation

* Corresponding Author's Email: ryadav@lko.amity.edu.

Introduction

Bioinformatics is the field of computational science that amalgamates different disciplines like biology, chemistry, physics, mathematics, computer science, information technology etc to solve various biological problem (Kour et al., 2021). Since decades bioinformatic has been used in drug discovery, drug target identification, understanding genes and function, system biology etc. Currently the power of bioinformatics and computational biology has been used in environmental sciences and to uncover environmental biology. This chapter discuss about the role of bioinformatics in environmental sustainability and sciences (Ricke et al., 2022).

Ecoinformatics is an amalgamation of the elucidated principles of bioinformatics and the data processing of the ecological information presented by nature. The ecosystem we live in is interdependent on a lot of previously built models, some being the food web equilibria and the stochastic theory of interdependence of various species spanning across the globe (Rabitz et al., 2022). Environmental changes reflect on the genotype and phenotype of biological characters that interact with the subsystem. This information is studied through created novel computational models, databases, and algorithmic codes for an in depth understanding of the interactions (Kumari and Kumar, 2021). Figure 1 shows the different branches of bioinformatics that are involved in different level of research like genomics, proteomics, system biology, metabolomics etc.

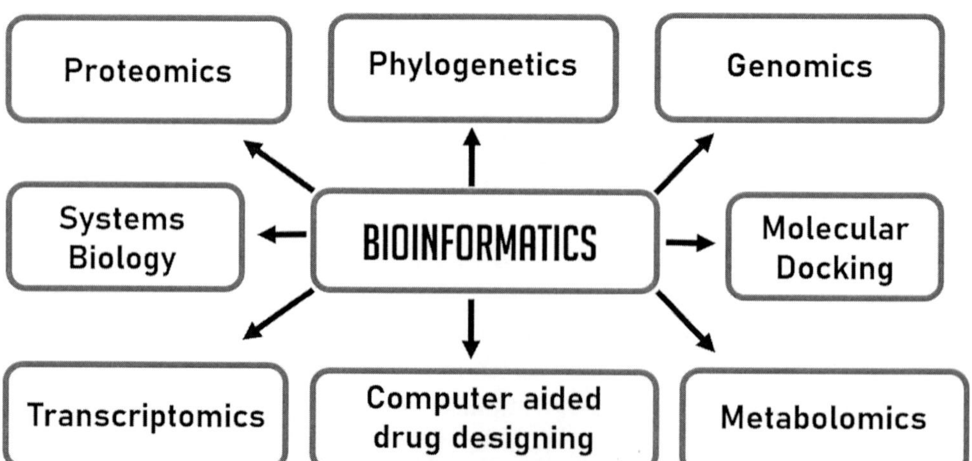

Figure 1. *Subdisciplines and branches of Bioinformatics*

Ecological informatics has been in circulation since the times of weather forecasting agencies and Seismographs were installed, notifying the people about ground motion and the atmospheric conditions through pattern recognition in fluctuating data (Effrosynidis and Arampatzis, 2021). This was given recognition in 2006 as a sub-branch of biodiversity informatics which held precedence. Considering the sheer number of observations collected daily about the ecological strata, observatories have been set up leading with NEON ((US National Ecological Observatory Network) and GLEON ((Global Lake Ecological Observatory Network) (Hounshell et al., 2021) to accumulate all collected information for a universal presentation to all scientists working in this area. These large- voluminous details are classified into numerous heads namely theoretical, empirical, statistical and in-silico depending upon the specific area of information (Jones et al., 2021).

As the world moves towards a more sustainable lifestyle with a huge emphasis on creating a better environmental culture more knowledge on this field is being amassed which just expands continuously till the biosphere level (Hanson et al., 2018). Acquisition of these facts are said to take place in a concentric – cyclical manner starting with methodical planning for collection, sorting out relevant material, testing tools and hypothesis for subsequent results, finally addressing the drawbacks to conclude the sequential cycle (Kominoski et al., 2018).

Ecology has very recently been conjoined with genomic and proteomic approaches. The leading repositories of environmental bioinformatic data freely available are listed below.

Bioinformatics and Environmental Sciences

Genomics

A genome is an organism is a whole set of DNAS, which includes all its genes and the three-dimensional, hierarchical structure organization they are organised into. It is the study of genomes structure, function, evolution and mapping is the focus of the interdisciplinary field of biology known as genomics (Formenti et al., 2022). By analysing the nucleotide sequence of genes, genomes, metagenomes, transcripts, transcriptomes, and meta transcriptomes, environmental genomics combines the diverse domains of knowledge about ancient and modern species and ecosystems. Environmental genomics offers a wealth of data on the taxonomy and diversity of living and extinct organisms, their phylogeny and evolutionary history, their capacity for adaptation and acclimatisation, their biology, their functional traits, and their interactions with the environment in both its biotic and abiotic dimensions (Bharagava et al., 2019). Table 1 shows the Bioinformatics Database for Environmental study along with information stored and URL

Table 1. *Databases used in Ecoinformatics with a short description and link.*

S.No	Database	Data Stored	URL
1.	Collective Molecular Activities of Useful Plants	Medicine obtained from plant sources	http://bidd.group/CMAUP/
2.	Vegbank	Vegetation records, community types and plant taxa.	http://vegbank.org/vegbank/index.jsp
3.	Barcode of Life Data System	Botanical DNA identifiers	http://v3.www.boldsystems.org/
4.	Net Seed	Large scale computational study of complex biological and ecological networks	http://borensteinlab.com/software_netseed.html
5.	Determinants of Radioresistance Database, or DetR DB	Radioresistant microbes and their molecular and genetic determinants	http://extremebiolab.com/detr-db/
6.	AgriGO	Agricultural associations and its Gene Ontology	http://systemsbiology.cau.edu.cn/agriGOv2/

Transcriptomics

The methods and technologies which are used to analyse the total number of RNA transcripts in an organism is known as transcriptomics. Transcription is the process through which an organism's information content is stored in its genome DNA (Harrill et al., 2021). Omics techniques provide the ability to track responses and changes in biological processes at the genome-scale, which is useful for anticipating potential negative impacts of environmental toxins. However, the high cost and absence of standardized toxicogenomic approaches prevent high throughput transcriptomics from being used in chemical assessment (Lamas et al., 2019).

Proteomics

The comprehensive study of proteomes is known as proteomics. A proteome is a collection of proteins produced by an organism, system, or biological setting. For example, a species; proteome (like that of Homo sapiens) or an organ (for example, the liver) (Guo et al., 2021). The study of variations in protein abundance and post-translational modifications, known as environmental proteomics, has emerged as a potent technique for developing theories about how the environment influences the biology of marine creatures. Previously unrecognised cellular impacts of environmental stresses, such as changes in temperature, osmotic, and anaerobic environments, are discovered by proteomics. In particular, those that emphasise post-translational alterations, proteomic investigations have helped characterise the biological impacts of pollutants and identified extensive and pollutant-specific collections of biomarkers (Carrascal et al., 2020).

Metagenomics

The study of genetic material extracted directly from environmental samples is known as metagenomics. Other names for the vast field include environmental genomics, economics, and community genomics. Metagenomics is the study of microbial communities, genomics without the use of culture (Chen et al., 2022). The analysis of DNA obtained directly from environmental samples provides a powerful tool for studying prokaryotes and viruses present in the nature. This technology takes into account the DNA of all microbes in a population. It not only identifies the microbial species present, but also provide information about the microorganism's functional roles and metabolic activities (Nowrotek et al., 2019).

Artificial Intelligence

The ARIES approach is a step-by-step assessment to classifying to classify theoretical literature data gathering into beneficial parts with high significant value to be added to a database. Every model/scenario entered under a similar head is segregated into quantitative and qualitative data which is learned by the system to produce maps like a Venn diagram with similarities. Probability routines are provisioned with the concluding results uploaded into a database to help study scenarios. These predictions are highly powerful as all similar interpretations are converted into a same format with duplicated information wiped out leaving only essential verdict (Nishant et al., 2020).

Software packages are continuously being updated in reference to ecology in the coding areas. There is a multiplex of questions when it comes to biodiversity under which evolution, patterns in phylogenetics are inherited but a dearth of solutions mainly because of unfulfilled formatting with unavailability of code (Yu et al., 2019). One package **Picante** is open-sourced under the R – CRAN repository using basic R plot functionality metrics to distinguish inputs into species diversity and clustering, evolutionary history through FAITH's algorithm, comparative analyses, environmental community regressions so on and so forth making it one of the most powerful and versatile packages available under the AI segment. It further incorporates a galaxy of statistical intelligence batch scripts like the Brownian probabilistic models; K means clustering parameterizations for a thorough testing this can be downloaded at http://picante.r-forge.r-project.org/ (Pearl et al., 2020).

Bioinformatics and Environment Sustainability

Plants and Herbal Drug Informatics

An area of herbal drug informatics cropped up as plants have been an integral resource in medicine and healing much before chemical formulations entered the system due to the redundancy and time-consuming nature of plant-based medicinal formulations. Systems biology in this area is a potential contributor to cutting costs and labour, hastening the remedial and study to unravel the potential benefits of plant-based medicine over synthetic formulations (Tanwar et al., 2018). Figure 2 shows the different approaches of Bioinformatics in Environmental Studies that is discussed further in chapter.

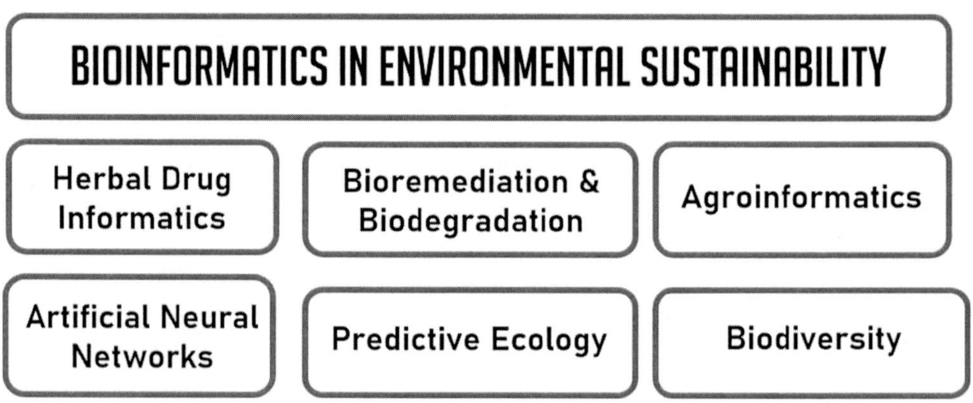

Figure 2. *Different approaches of Bioinformatics in Environmental Studies*

Plant utility is directly proportional to its therapeutic indexing. Algorithms are adept at predicting value due to which several FDA instructed drugs mapped out through in-silico approaches are in circulation - instance is the malarial artemisinin. E-science has taken the essence of long lost contemporary historical scientific findings of Traditional Chinese Medicine or the Ancient Ayurveda to quick validation thus back to recirculation in the modern era (Perrot et al., 2022).

Omics approaches in plants provide a more reliable method of judgement than phenotypical characteristics. In-vivo, the drugs are extracted followed by evaluation of activity of compounds in humans. Each flora has a dearth of information about their regulatory metabolic cycles, poisonous syntype differentiators, genomic profiles, metabolites which serve as better distinguishers than their height, shape, and other observable characteristics. This evidence can be gained by computational screening strategies such as DNA barcoding. Identification of hub genes are further supplemented by construction of an interconnected physiochemical properties network of identified active compounds in herbal drugs is an authentic validation strategy of certain medicinal plants (Fareed and Ali, 2022).

This myriad of documentation is now being stored in databases – a holistic place for access of medicinal ingredients.

- **Collective Molecular Activities of Useful Plants** (CMAUP) present in the URL http://bidd.group/CMAUP/ is a collection of utilitarian medicine obtained from plant sources. Target dataset gene ontologies defined by heatmaps of vegetation accumulated from 153 locations have been adequately categorized for mechanical research. Other than zonal division it also portrays its essential properties per species and diseases making it the topmost in research.
- **PlantCARE** https://bioinformatics.psb.ugent.be/webtools/plantcare/html/ is a small-scale transcriptomics webtool whose significance is storing the regulatory elements of a plant coupled with its domains and motif regions. A species entered as a query will output all the information about the classification, ranking, mitochondrial / chloroplast regions present in the system along with its referenced paper.
- **Barcode of Life Data System** http://v3.www.boldsystems.org/ under which the subset Medicinal Materials DNA Barcode Database (MMDBD) is a compendium of botanical DNA identifiers aka named barcodes for over 2000+ sequences available for standard identification. It has expanded to accommodate PSA and MSA accuracy methods along with alignment and designing of primers. Sequence databases are essential forecasters of undetermined – unspecified parts that can be researched once the genetic or proteome FASTA descriptor is known.

Bioremediation and Biodegradation

Bioinformatics is the amalgamation of basic biology and information technology which aims at cellular as well as molecular level of operation in modern biotechnology. Microbes depicts a miraculous range of contaminant degradation capacity which might efficiently and effectively rehabilitate the natural environment (Rodríguez et al., 2022). Bioremediation is a recent advanced technology that scrutinizes the microbial properties and their potentiality for biodegradation of harmful xenobiotic molecules. Bioinformatics needs the study of microbial genomics, proteomics, system biology, computational biology, phylogenetic trees, data mining and operations of major in – silico tools for determining and analyzing the structural properties and biodegradative pathway mechanism of these harmful xenobiotic compounds (Patil et al., 2022).

Environmental toxic waste has become a considerable global concern, by producing unwanted recalcitrant and xenobiotic molecules. A mixture of PAHs, xenobiotics molecules, chlorinated as well as nitro-aromatic molecules were observed to be eminently noxious, mutagenic and can carry cancer causing ability in living organisms (Breton-Deval et al., 2022). However, as a result of their diversification, versatileness and adaptability, numerous microbial species are considered to be the best contestants among all the various living organisms present in the world so as to re-mediate majority of the environment pollutants into legitimate biogeochemical cycle. These microbial species show a miraculous range of pollutant degradable potential which can efficiently refurbish natural environment (Trifi., 2020).

Bioremediation is defined process in which microbial species are stimulated to briskly degrade the unhealthy organic contaminants to environmentally intact levels of ground water, minerals, soil and sediments. Biodegradation is natural process of recycling wastes or degrading organic matter into smaller nutrients molecules which can be easily taken up by other living organisms (Ngara et al., 2022). The degradation process is being carried out by various microbial species such as bacteria, insects, worms, and fungi etc., by taking up macro-nutrients like carbon, phosphorous and nitrogen from the pollutant which on long run acclimatization and convert the noxious toxin compounds into non-hazardous eco-friendly compounds. Through composting, we expedite the natural occurring biodegradation and transform organic wastes into worthwhile resources. Wastewater treatment also stimulates the natural process of biodegradation and degrades the organic matter into nutrients or non-hazardous compounds so that when it's released in the environment it does not cause any environmental pollution (Kour et al., 2021). Through bioremediation, numerous microbial species are being used in cleaning of oil spills and various other kinds of organic pollution.

The fields of Biodegradation and Bioremediation provides numerous interesting and uncharted possibilities from the bioinformatics viewpoint. They require to incorporate a large amount of information from various sources like chemical structural properties and the reactivity of the organic molecules, sequences, structure, and function of the targeted proteins, comparative genomics analyses and environmental biology etc (Rawat and Rangarajan, 2019). Bioinformatics contributes databases for microarrays, identification of specific genes and microbial degradation pathways of xenobiotics.

Nowadays, researchers have the potential to cultivate and grow microbes which are significant in bioremediation and can analyse their physiology by applying a unification of gene-enabled experimental data and various modelling techniques (Bibi et al., 2019). During mineralization, transcriptional structures and their functional expression have been studied by using high-throughput sequencing and transcriptomics approach with microarrays. Generally, the studied transcripts have no potential to perform any physiological functions; so, they must be converted and translated into proteins with meaningful functional expression levels. The protein obtained are identified by applying the current proteomics technique using robust 2-dimensional polyacrylamide gel electrophoresis (2- DE). While the establishment of the structural and functional proteomics study, the current approaches in mass-spec (MS) and protein micro-arrays play a pivotal role in the proteomic techniques. Exploring and studying the differential expression for an immense assortment of proteins and screening the entire sequence of the genome for proteins that associate with specific mineralization regulatory pathways would aids us to achieve insights of bioremediation and biodegradation (Srinivasan and Sadasivam, 2018).

Agroinformatics Sustainability

Agricultural bioinformatics is one of the current and rapidly growing disciplines of live science which exploits in-silico techniques to responsive biological sciences. The upsurge in industrial agricultural codes had led to gene destruction and owing to the variations caused in climatic conditions. India has mostly concentrated on technological advancement, but there aren't enough effective governmental measures to cause a change away from conventional technology. The National System of Innovation framework, which may be mapped for research purposes, illustrates the condition of agriculture bioinformatics as it stands right now (Paul et al., 2019). The current state of agriculture bioinformatics is shown in the National System of Innovation framework, which may be mapped for research purposes. The framework needs networks of stakeholders with a range of skills to produce results that can meet both the present and future food demand and identify problems. There are some research institutions in India that place a strong emphasis on crop improvement through gene tagging, gene editing, genome sequencing, and other techniques for using them to increase productivity, resistance to diseases, and other biotic and abiotic changes, which are a threat to plants' ability to retain their properties in changing climatic conditions and with changing nature of soil content and soil texture (Wadghane, 2022). Technology spreads through a process called adoption and diffusion that starts with its introduction and ends with its ubiquity. The main players in the field of Agro bioinformatics in India are public sponsored research institutes and universities. They have been instrumental in helping themselves collaborate with other research organisations. Weak ties have continued to play a significant role in changing current laws so that institutions can gradually build up a technological trajectory. Such ties are unable to affect change in socio-technical systems because they obstruct the exchange of sophisticated knowledge, which demonstrates a lack of trust (Ciurko et al., 2022).

Sustainable agricultural production is crucial problem in response to worldwide climate change and escalating population growth rate. Moreover, current raise in the demand of biofuel crops has put together a new sector in the market for agricultural commodities. One potential resolution is to boost plant crop yield by designing and creating plants on the basis of a good molecular understanding of numerous gene function and also on their regulatory mechanisms incorporated in stress tolerance and plant growth. Recent advancement in plant genomics has permitted us to discover, identify and isolate significant genes of interest and to examine functions that stimulates higher plant yields and tolerance to various environmental stresses (Paul et al., 2020).

Recently, several methods for identifying, characterising, and analysing the function of plant genes involved in signalling, triggering, and responding to biotic and abiotic stimuli have been proposed. Understanding the interactions between plants and diseases at the gene and genome level requires in silico biology (Reddy et al., 2012).

AgriGO – It is a Gene Ontology toolkit especially made for agricultural associations present with four customizable attributes for a variety of deliverables accessible at http://systemsbiology.cau.edu.cn/agriGOv2/. Its architecture is designed for conducting statistical and mathematical modelling plots along with graphical visualization codes to note the p-value distribution as well as comparative interpretation tools namely SEA, PAGE, transfer BLAST.

VegBank – Vegbank is a platform encompassing all the metadata of the plot documentation of all taxa and vegetation per location per plot available in the USA for easy retrieval. It hides the floral composition including soil composition statistics by the usage of field observations. It covers the entire basal area of plant communities with a unique feature to contribute plots along with normalizing which is then reflected in the map. Available at the URL http://vegbank.org//index.jsp it encompasses other unique variable components: slope of plot, its elevation, time span of plantation, types of crops at a particular season which would bloom amongst other making it a very wholesome essential directory for farming.

Conclusion

As long as we live on the planet, we will be conducting numerous mishaps while utilizing its resources. Thus, there will always be a created need for informatic studies to minimize disastrous impacts. Explorations on ozone depletion, escalating concentrations of CO_2, appalling high margins of pollution in the air, ocean, land desperately call for long term maintenance of the environment we live in. Catastrophes are occurring at an alarming rate owing to global warming. All of this generates a humongous collection of archival data of earthly observations which is now possible to analyse through Web maturity and expansion. Ecoinformatics, the entire discipline is making it easier to identify indicators, approaches remotely hence rectifying the problematic areas where we are bound to cause errors. It is also supportive in model building to continue expanding discoveries in the environmental sciences benefitting nature forthwith in tandem humans too in terms of medical breakthroughs, genetically modified food, agriculture, age old herbal practices, waste disposals on top of other reasonable areas off ecological studies without being in the actual location.

Despite all the advancements there is a lot of impeding strategies still being applied to environmental sciences and its allied sides. We are still a far cry from understanding epidemiology transmission in humans when related to the natural world. Inspecting any environmental data is highly heterogenous as there is still a lot of undiscovered aspects to this scientific discipline. Frequency of any characteristic trait is highly variable at different points of time throughout the lifespan of any eco related species making any documentation and storage a tedious mandatory process. Going from the biosphere to singular gene study is challenging hence delving into ecological informatics surveys are the new hot topic to be tackled for ease of access to the public for benefit of the humankind.

Disclaimer

None

References

Bharagava RN, Purchase D, Saxena G, Mulla SI. Applications of metagenomics in microbial bioremediation of pollutants: from genomics to environmental cleanup. *In Microbial diversity in the genomic era* (2019) 459-477

Bibi T, Manzoor S, Azhar S, Saeed M, Asif Raheem M. Role of bioremediation in environment. *Middle East Journal of Applied Science & Technology* (2019) 2:54-9.

Breton-Deval L, Sanchez-Reyes A, Sanchez-Flores A, Juárez K, Salinas-Peralta I, Mussali-Galante P. Functional analysis of a polluted river microbiome reveals a metabolic potential for bioremediation. *Microorganisms* (2020) 8(4):554.

Carrascal M, Abian J, Ginebreda A, Barceló D. Discovery of large molecules as new biomarkers in wastewater using environmental proteomics and suitable polymer probes. *Science of The Total Environment* (2020) 747:141145.

Chen G, Bai R, Zhang Y, Zhao B, Xiao Y. Application of metagenomics to biological wastewater treatment. *Science of The Total Environment* (2022) 807:150737.

Ciurko D, Czyżnikowska Ż, Kancelista A, Łaba W, Janek T. Sustainable Production of Biosurfactant from Agro-Industrial Oil Wastes by Bacillus subtilis and Its Potential Application as Antioxidant and ACE Inhibitor. *International Journal of Molecular Sciences* (2022) 23(18):10824.

Effrosynidis D, Arampatzis A. An evaluation of feature selection methods for environmental data. *Ecological Informatics*. (2021) 61:101224.

Fareed MM, Ali MM. The Revelation and Therapeutic Role of Medicinal Phytochemicals in the Treatment of Cancer: A Brief Review. *Computational Intelligence in Oncology* (2022) 335-47.

Formenti G, Theissinger K, Fernandes C, Bista I, Bombarely A, Bleidorn C, Ciofi C, Crottini A, Godoy JA, Höglund J, Malukiewicz J. The era of reference genomes in conservation genomics. *Trends in ecology & evolution* (2022)

Guo H, Wang L, Deng Y, Ye J. Novel perspectives of environmental proteomics. *Science of The Total Environment* (2021) 788:147588.

Hanson PC, Weathers KC, Dugan HA, Gries C. The global lake ecological observatory network. *In Ecological Informatics* (2018) 415-433.

Harrill JA, Everett LJ, Haggard DE, Sheffield T, Bundy JL, Willis CM, Thomas RS, Shah I, Judson RS. High-throughput transcriptomics platform for screening environmental chemicals. *Toxicological Sciences* (2021) 181(1):68-89.

Hounshell AG, Farrell KJ, Carey CC. Macrosystems EDDIE Teaching Modules Increase Students' Ability to Define, Interpret, and Apply Concepts in Macrosystems Ecology. *Education Sciences*. (2021)11(8):382.

Jones JA, Groffman PM, Blair J, Davis FW, Dugan H, Euskirchen EE, Frey SD, Harms TK, Hinckley E, Kosmala M, Loberg S. Synergies among environmental science research and monitoring networks: a research agenda. *Earth's Future* (2021) 9(3):e2020EF001631.

Kominoski JS, Gaiser EE, Baer SG. Advancing theories of ecosystem development through long-term ecological research. *BioScience*. (2018) 68(8):554-62.

Kour D, Kaur T, Devi R, Yadav A, Singh M, Joshi D, Singh J, Suyal DC, Kumar A, Rajput VD, Yadav AN. Beneficial microbiomes for bioremediation of diverse contaminated environments for environmental sustainability: present status and future challenges. *Environmental Science and Pollution Research*. (2021) 28(20):24917-39.

Kour D, Kaur T, Devi R, Yadav A, Singh M, Joshi D, Singh J, Suyal DC, Kumar A, Rajput VD, Yadav AN. Beneficial microbiomes for bioremediation of diverse contaminated environments for environmental sustainability: present status and future challenges. *Environmental Science and Pollution Research* (2021) 28(20):24917-39.

Kumari P, Kumar Y. Bioinformatics and computational tools in bioremediation and biodegradation of environmental pollutants. *In Bioremediation for environmental sustainability* (2021) 421-444.

Lamas A, Regal P, Vázquez B, Miranda JM, Franco CM, Cepeda A. Transcriptomics: A powerful tool to evaluate the behavior of foodborne pathogens in the food production chain. *Food Research International* (2019) 125:108543.

Ngara TR, Zeng P, Zhang H. mibPOPdb: An online database for microbial biodegradation of persistent organic pollutants. *iMeta* (2022) e45.

Nishant R, Kennedy M, Corbett J. Artificial intelligence for sustainability: Challenges, opportunities, and a research agenda. *International Journal of Information Management (*2020) 53:102104.

Nowrotek M, Jałowiecki Ł, Harnisz M, Płaza GA. Culturomics and metagenomics: In understanding of environmental resistome. *Frontiers of Environmental Science & Engineering (*2019) 13(3):1-2.

Patil A, Yesankar P, Bhanse P, Maitreya A, Kapley A, Qureshi A. Omics Perspective: Molecular Blueprint for Agrochemical Bioremediation Process in the Environment. *In Agrochemicals in Soil and Environment* (2022) 585-608.

Paul P, Aithal PS, Ripu Ranjan Sinha RR, Saavedra M R, Aremu PS. Agro Informatics with its Various Attributes and Emergence: Emphasizing Potentiality as a Specialization in Agricultural Sciences—A Policy Framework. *IRA International Journal of Applied Sciences (*2019) 14(4):34-44.

Paul P, Ripu Ranjan Sinha RR, Aithal PS, Saavedra M R, Aremu PS. Agro Informatics Emphasizing Its Potentiality as a Full-Fledged Degree Programs: International Context & Indian Potentialities. *Asian Journal of Computer Science and Technology*. (2020) 9(1):19-26.

Pearl H, Ryan T, Howard M, Shimizu Y, Shapcott A. DNA Barcoding to Enhance Conservation of Sunshine Coast Heathlands. *Diversity* (2022) 14(6):436.

Perrot J, Bennetau-Pelissero C, Miremont-Salamé G, Petitet F, Cluzet S, Peyrouzet H, Letinier L. Herb-drug interaction resources: The PHYDGI database. *Phytomedicine Plus*. (2022) 2(3):100324.

Rabitz F, Feist M, Honegger M, Horton J, Jinnah S, Reynolds J. A preliminary framework for understanding the governance of novel environmental technologies: Ambiguity, indeterminateness and drift. Earth System Governance. (2022) 12:100134.

Rawat M, Rangarajan S. Omics approaches for elucidating molecular mechanisms of microbial bioremediation. *In Smart Bioremediation Technologies* (2019) 191-203.

Reddy KA, Prasad JR, Kumar KP. A Study on Agriculture Informatics in Indian Context. *International Journal of Computer Applications*. (2012) 55(12).

Ricke SC, Dittoe DK, Tarcin AA, Rothrock Jr MJ. Communicating the utility of the microbiome and bioinformatics to small flock poultry producers. *Poultry Science*. (2022) 101(5):101788.

Rodríguez A, Castrejón-Godínez ML, Salazar-Bustamante E, Gama-Martínez Y, Sánchez-Salinas E, Mussali-Galante P, Tovar-Sánchez E, Ortiz-Hernández M. Omics approaches to pesticide biodegradation. *Current Microbiology* (2020) 77(4):545-63.

Srinivasan S, Sadasivam SK. Exploring docking and aerobic-microaerophilic biodegradation of textile azo dye by bacterial systems. *Journal of water process engineering* (2018) 22:180-91.

Tanwar A, Zaidi AA, Bhardwaj M, Rathore A, Chakotiya AS, Sharma N, Sharma D, Kulshrestha S, Chawla R, Basu M, Arora R. Herbal informatics approach for the selection of natural compounds targeting diabetes mellitus (2018)

Trifi H, Najjari A, Achouak W, Barakat M, Ghedira K, Mrad F, Saidi M, Sghaier H. Metataxonomics of Tunisian phosphogypsum based on five bioinformatics pipelines: Insights for bioremediation. *Genomics* (2020) 112(1)981-9.

Wadghane R. Sustainability Management Status of Agro-Ecosystems: A Case Study of Sugarcane Farmers in Shevgaon and Paithan (Sub-Districts) of Maharashtra, India. *Agricultural Research* (2022) 21:1-0.

Yu KH, Zhang Y, Li D, Montenegro-Marin CE, Kumar PM. Environmental planning based on reduce, reuse, recycle and recover using artificial intelligence. *Environmental Impact Asse*

Chapter 8

Environmental Informatics Current Research and Future Prospects

Wayez Naqvi and Ruchi Yadav*

Amity Institute of Biotechnology, Amity University Uttar Pradesh, Lucknow Campus, Lucknow, UP, India

Abstract

Environmental informatics (EI) is branch of environmental sciences that deals with the metadata information related to environment. EI deals with the software, tools, and databases to support sustainable environment. EI includes the information metadata information related to environment use for research, protection and environmental studied. This chapter introduces EI and its application in sustainable environment, current research, application, and future prospects. Environmental informatics studies environmental challenges that can be both discrete and continuous in space or time using huge multidimensional, complex information. These datasets and the metadata that goes with them can be maintained via query able databases. Remote access to dynamic subsets of environmental information is provided through geospatial Web application programming interfaces (APIs). Data is citable thanks to persistent identifiers. The storage-computing trade-off has shifted substantially in favour of shifting computations to the data. Provenance information aids in determining the dependability and trustworthiness of a data asset. Long-term records were used to reveal rising atmospheric CO2, the Antarctic ozone hole, and Gulf Stream warm-core rings. Similar work is being done to map evapotranspiration and snow water equivalent. Data (particularly data obtained operationally) drives hypothesis creation in these "fourth paradigm" challenges. Making data available necessitates the development of new discovery tools as well as data-sharing regulations. Cloud computing and array-friendly databases will aid in data processing. Citizen scientists will be able to collect environmental data thanks to ubiquitous location sensing and geotagging.

Keywords: Array-Friendly Databases, Cloud Computing, Geotagging, Provenance.

---

* Corresponding Author's Email: ryadav@lko.amity.edu.

Introduction

Environmental Informatics is an interdisciplinary academic area that studies the environment and seeks answers to environmental problems by combining physics, biology, and geography. Environmental scientists study the earth's physical, chemical, biological, and geological processes to better understand how issues like alternative energy systems, pollution control and mitigation, natural resource management, and the consequences of global warming and climate change influence and alter the earth's natural systems and processes (Paul and Aithal, 2020). Physical, chemical, and biological processes usually invariably interact in environmental challenges. Environmental scientists investigate environmental issues from a systems approach. A competent environmental scientist must be able to link location and temporal relationships, as well as do quantitative analysis. Informatics is the study of the structure, characteristics, and exchange of information, as well as the means of storing or processing information. Environmental informatics is defined as the use of data science to environmental issues (Paul et al., 2020).

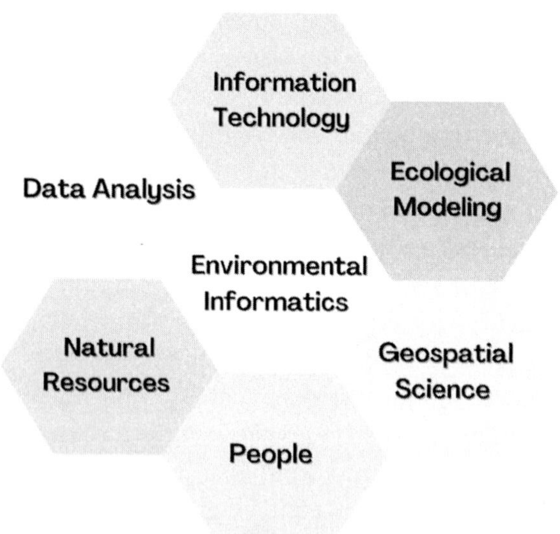

Figure 1. *This diagram depicts the numerous areas that have converged to form Environmental Informatics.*

Computational paradigms and techniques that focus primarily on the collection, management, exploitation, sharing, and preservation of environmental information are used in the framework of data science. Figure 1 shows the different branches of environmental informatics that includes Ecological modeling, Geospatial science, data analysis etc. Data science, as described, is the subset of information science that focuses on huge multidimensional datasets with complicated structures (Nassour et al., 2020). By environmental information, emphasis is given to data regarding the status of the Earth's biosphere (and adjacent spheres) and the processes that effect it. Environmental information varies from generic information in that its dimensions always involve space and time, and its accurate interpretation necessitates extensive scientific background (i.e., metadata) (Cioruta and Coman, 2019)

Characteristics of Environmental Information

Spatiotemporal Information

Environmental information defines traits and events that have distinct spatial manifestations and are not evenly distributed throughout the Earth's biosphere and its immediate surroundings. As a result, environmental informatics must manage both the spatial representation of environmental occurrences and the Earth's surface locations linked with these phenomena (Chowdhury, 2019).

Earth Coordinate System

Locations on the Earth's surface can be described directly using spatial coordinates or implicitly using names or limitations. Although maintaining names is common in informatics, managing geographical coordinates is unique to environmental informatics. Geographic and/or projected coordinates are used to specify locations on the Earth's surface. Geographic coordinates are angular coordinates (the familiar latitude and longitude) that are referenced to an ellipsoidal approximation of Earth's true shape (the geoid), whose minor axis is parallel to Earth's, and whose surface bears a specific relationship to a reference location on Earth's surface. A datum is formed by the combination of an ellipsoid and a reference location (Wardani and Wahono, 2020).

Although the World Geodetic System 1984 (WGS 84) has become the global standard for geographic coordinates, historical (and some current) environmental information employs geographic coordinates related to alternate datums. To compare geographic coordinates correctly, they must first be translated to a similar datum. Most geographic processing software includes these modifications, however environmental data is commonly managed without explicitly monitoring the reference datum. At large map sizes, failure to account for the datums of geographic coordinates is a typical source of positional mistakes (Zhang et al., 2019).

Projected coordinates are rectangular coordinates that arise from projecting geographic coordinates onto a developable surface (plane, cone, cylinder, or other).

The type of surface, its connection to the datum, and the projection geometry all define a projection (orthographic, stereographic, etc.). Because an ellipsoid cannot be developed, any projection of the entire Earth will be discontinuous. More crucially, all projections distort distances (thus, areas) and/or angles (hence, shapes) in predictable ways, albeit these distortions can be traded off against each other and reduced for specified fractions of the Earth's surface. Equal-area projections, which retain area connections at the price of form accuracy, are commonly employed for environmental assessments because they allow distances and areas to be computed using basic Euclidean (rather than elliptical trigonometry) procedures (Hori et al., 2021).

Spatial Information Representation

Spatial phenomena can be classified as either discrete or continuous. Discrete phenomena are linked to discrete things, which in turn are linked to specific locations. "Diameter at breast height," for example, is connected with a certain tree and makes no sense in the absence of a tree. Discrete objects are traditionally represented in environmental informatics as having basic Euclidean spatial representations based on points and their composition into lines, polygons, circles, and so on (Hao et al., 2021).

A full representation for discrete spatial phenomena must thus integrate geometric information describing the position and extent of the phenomenon with nonspatial information describing the value of the phenomenon. This hybrid spatial (geometry)/nonspatial (attribute) representation is known as a georelational model. The "shapefile" data format and SQL database management systems (DBMSs) that support the SQL/MM (Multimedia and Application Packages) spatial data types are common instantiations of the georelational paradigm (Zhao, 2018).

Continuous phenomena (also known as spatial fields) are functions defined across all conceivable spatial places (for example, surface temperature is a spatially continuous phenomenon). Although continuous phenomena can be thought of as functions, the exact functions that produce the phenomena are frequently unknown. As a result, the most frequent way to describe a spatial field is to sample the observed values of the field. Although these samples can be represented as discrete phenomena, they are more commonly represented as a multidimensional array of samples, with some dimensions representing spatial (and possibly temporal) coordinates and the remaining dimensions representing the dimensionality of a sample at a single location (Imani, 2020).

Because most programming environments allow multidimensional arrays as fundamental data structures, an array representation facilitates the computational implementation of a field-based analysis. Furthermore, arrays efficiently encode positions, which may be computed from the array's corner location, orientation, and cell size, rather than being explicitly recorded with each sample (Xu et al., 2020).

Temporal Information

Managing the time dimension of information is common in informatics, but it is especially important in environmental informatics due to the wide range of timeframes involved. The concerns addressed by temporal data can be generically classified as periodicity, duration, and synthesis. Many environmental events occur on a regular basis. Some, such as diurnal, tidal, seasonal, and yearly variations, are caused directly by interactions between the Earth, Sun, and Moon. Others are influenced by more regularly occurring external processes, such as the 11-year sunspot cycle. Other periodic occurrences, such as the El Nino-Southern Oscillation and the Madden-Julian Oscillation, are caused by the redistribution of differential global heating by the land-ocean-atmosphere system. Rather of referring to an exact time base, these secondary occurrences are frequently referred to by name. All these different types of periodicities must be considered by an environmental information management system. Environmental events can last for long periods of time. A tornado can be over in minutes; a snowpack can melt completely in a year; and a glacier can progress and recede over thousands of years (Jalal et al., 2020).

Although, all else being equal, the temporal resolution at which a phenomenon must be sampled coarsens as the time of the phenomenon rises, environmental investigations frequently must deal with predefined (rather than optimum) sampling intervals. The Landsat satellites, for example, have a recurrence interval of 16 days; phenomena that fluctuate more often than this would not be fully captured. Furthermore, individual samples of environmental phenomena may be aggregated due to sensor properties like as binning or dwell length (Wei et al., 2021).

To minimize data sizes or exclude undesired short-term phenomena, many environmental datasets are generated as temporal composites (e.g., average, minimum, maximum) of several shorter-term observations. As a result, the administration of an environmental dataset may need keeping track of many temporal attributes: acquisition time, integration/dwell time, sampling interval, and aggregation time span (Chen.. 2020).

Metadata models and standards

The term metadata has several connotations in informatics, ranging from the limited "database schema" to the broad "data about data." Preference is given to the role-based definition: Metadata are data items that offer context for the understanding and usage of other data objects in a specific study. Indeed, metadata and context are virtually equivalent in this situation. The context necessary to comprehend environmental data is frequently complicated. As previously stated, all environmental information contains spatiotemporal and type features that constitute nontrivial context. As a result, metadata management has become an important component of environmental informatics, showing itself in two ways: standards and implementations (Anthony A. Leiserowitz, 2006). Figure 2 shows the different functions of data management and different levels of data collection, storage, management, data models etc (Khaiter and Erechtchoukova, 2019).

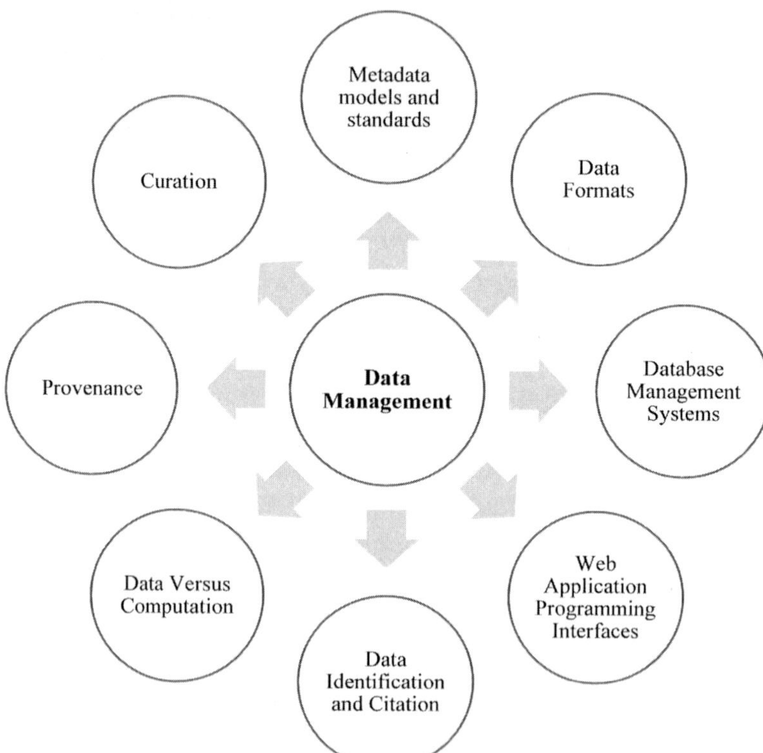

Figure 2. *Data Management and its various features*

Environmental metadata standards provide organized direction to the types of information that are most frequently required to comprehend environmental data. The Content Standard for Digital Geospatial Metadata (CSDGM), first promulgated in 1998 by the US Federal Geographic Data Committee in response to a presidential executive order mandating the publication of most spatial information held by

US government agencies, is one thread of environmental metadata standardization (Gordon and Habermann., 2018).

The CSDGM is a content standard, which means it describes the types of information needed to fully characterize an environmental dataset and their hierarchical connection but does not specify how the material should be recorded. As a result, several formats, including as structured text, HTML, and XML, have become widely used (Hardisty et al., 2019).

Some of this information may be extracted from the data itself (for example, the bounding box for the data's spatial extent), but the majority must be given independently, either by the environment in which the data is formed or by a human participating in the data's generation or maintenance. This is accommodated by environmental software, which includes metadata editors (for human input) as well as automatic information extraction. The CSDGM had a significant impact on ISO 19115, the current worldwide standard for geospatial metadata. Both standards, which are significant for environmental informatics, are based on a "core plus extensions" approach, in which the basic standard describes common features and extensions specify elements relevant for specific application areas, such as remote sensing (Michener, 2018).

Data Formats

Environmental information is complicated, requiring consideration of coordinate systems, spatial and temporal representations, and linkage with similarly complex metadata. Various packaging systems, or formats, are widely used to connect these ideas and allow a single item to include enough context to allow its successful usage.

The shapefile format is a straightforward format for combining homogenous geographic information with a single set of scalar attribute tuples. An XML-based format, such the Keyhole Markup Language (KML), on the other hand, may enable heterogeneous features, various attribute sets, and hierarchical nesting. Because of Google's sponsorship, KML has gained widespread acceptance. A KML document may include style information (instructions on how the data should be presented), making it an effective format for publishing feature information; the KML document can be viewed as a digital map that includes the data on which it is based (Ghosh and Maiti, 2021).

GeoTIFF is a common raster (field) data format. GeoTIFF adds metadata for the raster's map project and ground coordinates to the ubiquitous TIFF format. GeoTIFF metadata "tags" are harmlessly disregarded by software that does not comprehend them due to the flexible architecture of TIFF. Any TIFF-capable imaging program may so readily output a GeoTIFF file. Complicated data structures and rich information can be bundled into a single file or data stream in more complex formats. The Network Common Data Form (netCDF) format is widely used in atmospheric and ocean research groups. NetCDF offers multidimensional arrays in addition to user-defined embedded information, making it an excellent alternative for continuous phenomena (Jaspersen et al., 2018).

The Open Geospatial Consortium (OGC) standard Geography Markup Language (GML) is an extremely comprehensive XML dialect capable of modelling any geospatial construct. GML supports features and coverages (fields), as well as specifications for spatial and temporal coordinate systems, styling information, and dynamic features. The most detailed official base map (Master map) of the United Kingdom has been entirely coded in GML, demonstrating GML's capability (Wöllauer et al., 2021).

Database Management Systems

In environmental science, the term database can refer to a variety of things. It is frequently used in a broad sense to refer to any collection of information that can be treated (for example, a set of files containing all of the raw data from a set of field instruments). We define DBMS more narrowly as a system that integrates information management and access according to common schemata. The relational data model is implemented in the DBMSs most used by environmental scientists: Classes are tables, objects are rows, properties are columns, and comparing property values implies relationships between classes (as opposed to being represented explicitly by links) (Porter, 2018).

As a management mechanism for environmental information, DBMSs are frequently contrasted with "flat files." Every multiuser computing environment includes a hierarchical file system that allows files to be stored in (possibly nested) folders (also called directories). Querying data in a flat file entail traversing directories to locate the file and then reading the file to learn about its contents. All information stored in a DBMS, on the other hand, can be retrieved by queries that specify the source tables and the rows to be retrieved from them, based on whether the column values satisfy the logical constraints specified in the query. In effect, the DBMS includes the common operations of combining and sub setting, which can greatly simplify the construction of environmental analyses (Afanasyev et al., 2019).

- DBMS software has become both widely available and reasonably priced. Open-source DBMS servers, such as MySQL and PostgreSQL, are now capable of competing with their proprietary counterparts.

- Database design and implementation skills are becoming more common, allowing small scientific teams to create and maintain their own databases.
- Scientific computing tools, such as MATLAB and R, now routinely connect to DBMSs, making databases potentially more appealing as a storage mechanism for query able information than flat files.
- DBMSs are increasingly supporting the extended data types and operators required to effectively support environmental information, such as geometry, arrays, and user-defined functions. Almost all environmental information can now be stored in a DBMS.

Web Application Programming Interfaces

The World Wide Web's explosive growth has transformed all aspects of informatics. In terms of environmental informatics, we believe the Web has had a significant impact by enabling or simplifying programmatic access to remote resources through Web-based application programming interfaces (APIs) that enable environmental information's particular qualities (space, time, type, and others). The Distributed Ocean Data System (DODS) and its successor, the Open-Source Project for a Network Data Access Protocol (OPeNDAP), are leading efforts to support Web-based programmatic access to environmental data (Tandule et al., 2022).

OPeNDAP is made up of a network-based ("wire") format for transmitting structured data, a collection of ancillary formats for transmitting metadata, and a collection of messages that a client can use to request data or metadata from a server. OPeNDAP includes reference server implementations that support common environmental data formats (e.g., HDF, netCDF) as well as client implementations for common analysis environments (e.g., MATLAB). OPeNDAP, like the Web, was originally designed to provide end-to-end support for a specific distributed community (in the Web's case, high-energy physicists; in OPeNDAP's case, oceanographers), and, like the Web, OPeNDAP was designed to provide end-to-end support for a specific distributed community (in the Web's case, oceanographers). It now has a vibrant multidisciplinary user community that contributes alternative server implementations as well as support for new clients and data formats (Gan et al., 2020).

Data Identification and Citation

In many environmental domains, data collection and assembly are significant resource-intensive activities in and of themselves. To publish data, two challenges must be addressed: granularity and identification. After establishing a publication granularity, the published granules must be identified in a form that allows permanent citation. The digital object identifier (DOI) is one potential system for this (Martin et al., 2021).

Data Versus Computation

Computing systems are distinguished by their storage capacity, computational power, and data bandwidth. The end-to-end throughput of a data storage and transmission system, including any network or request latencies in addition to raw channel bandwidth, is referred to as bandwidth. Indeed, the amount to which any of these limitations was dominant at any given period has substantially controlled the evolution of computer system architecture. In terms of environmental data, the most major decisions compelled by these limits are caching and recreation. The content of environmental datasets ranges from simply observational data to fully synthetic data derived from models. Because observational data cannot be recreated, an analysis incorporating observational data requires direct access to or a copy of that data. Synthetic data, on the other hand, can theoretically be produced by rerunning the source model with the same settings and inputs (Pipattanasomporn et al., 2020).

When storage space is more constrained than bandwidth for datasets that cannot be regenerated, the most cost-effective option to access enormous datasets is to reload them from remote or off-line versions. When bandwidth is more limited than store space, the most cost-effective option to access massive datasets is to keep local online copies. This is the caching approach used by Web browsers to preserve a local copy of a previously visited Web page that can be referenced if the page is revisited, avoiding the need to submit another network request for the page. By local, we imply storage with the highest available bandwidth—in computing, closeness is measured by access speed rather than physical distance (Nag and Cummins, 2021).

Provenance

The provenance of a digital thing is defined as "the process that resulted in" that object. Provenance is essential for assessing an object's dependability and suitability for purpose, as well as diagnosing and establishing the effects of faulty data or processing. Existing environmental metadata standards all offer free-text provenance representations (essentially, a narrative pedigree of an object) (Dundore-Arias, 2020).

Curation

Digital curation is an active care of information that includes both preservation (ensuring that content does not vanish) and accessibility. Even though traditional publishing has created a medium, acid-free paper, that can last for centuries, no popular digital storage medium has yet to last more than 30 years. As online storage becomes significantly less expensive, previously competitive off-line storage systems may be phased out, rendering linked offline data unavailable. Part of the digital curation dilemma is ensuring that material remains understandable in the digital era. Because environmental data is so complicated, a significant quantity of metadata is frequently necessary to fully interpret it. Traditional curation, which is geared toward conserving language and objects, frequently relies on metadata that is deeply embedded in culture (Beretta et al., 2021).

Data Science Examples

Six noteworthy breakthroughs in data-driven discovery are highlighted, in which valuable environmental knowledge was uncovered or clarified. Atmospheric science, oceanography, sustainability, ecology, and hydrology are among examples. The first three—atmospheric carbon dioxide (CO2) concentration, ozone depletion, and the structure of the Gulf Stream—show how data analysis led to crucial and startling discoveries about our environment. The latter three—indicators of sustainability in the context of climate change and geographical distributions of evapotranspiration and snow—illustrate difficulties that are only now being addressed and necessitate the use of concepts and techniques from environmental informatics.

Rising Concentrations of Atmospheric Carbon Dioxide

The Keeling Curve, possibly the most known graph in environmental science, depicts Charles David Keeling's 1958 record of atmospheric CO2 concentrations on Mauna Loa in Hawaii. Within a few years, continuous data from Mauna Loa revealed that CO2 concentrations varied seasonally in response to photosynthesis and respiration, but that they clearly exhibited a rising trend, in tandem with readings from the South Pole. If CO2 had been detected using worldwide periodic surveys instead, evidence for the trend would have come decades later. The story demonstrates the value of data-driven discovery, the relevance of precise measurements, and the significance of learning from long-term continuous Earth monitoring (Appleyard et al., 2021).

The Antarctic Ozone Hole

The narrative of ozone depletion shows how laboratory tests, surface observations, and extensive satellite data led to the realization of anthropogenic halogen-containing chemicals' ozone-depleting activity. F. Sherwood Rowland and his postdoctoral fellow Mario J. Molina released laboratory research in 1974 indicating that chlorofluorocarbons, a type of halocarbon, catalytically inhibit stratospheric ozone production in a UV light environment. 1 The identical series of chemical reactions was identified by Stolarski and Cicerone. Rowland and Molina projected that continued ozone depletion would result from the release of chlorofluorocarbons into the atmosphere. Although halocarbons were quickly phased out as propellants in aerosol sprays, the industry battled their prohibition as refrigerants and other applications (Dhomse et al., 2019).

The breakdown of chlorofluorocarbons produces relatively innocuous hydrochloric acid, which reacts on the surface of ice crystals in polar stratospheric clouds, which were also identified by satellite measurements, to produce chlorine monoxide, which catalytically destroys ozone. Satellite measurements also confirmed the presence of bromine monoxide, which is engaged in reactions that deplete stratospheric ozone even further.

Satellite measurements continue to follow the size and depth of the Antarctic ozone hole, as well as the more subtle, but potentially catastrophic, ozone losses over densely populated areas. Recent satellite observations suggest a reduction in halocarbons and the apparent start of an ozone recovery, enhancing confidence that the Montreal Protocol is succeeding (Spolaor et al., 2021).

The Gulf Stream and Its Warm-Core Rings

Observations from ships, buoys, and floating sensors in the open ocean cannot resolve the temporal and spatial variability required to comprehend the ocean's dynamic nature. Satellite observations supplied the first globally available data. The proper categorization of the Gulf Stream, for which earlier intense in situ studies had failed to disclose the correct structure, is an example of data-driven discovery. Satellite measurements of sea-surface temperature yielded information on the structure of the circulation as well as precise information about heat movement in the ocean (Gangopadhyay et al., 2019).

Oceanographers suspected that the Gulf Stream was made up of several currents based on ship readings, and ship data could not distinguish between different interpretations. However, in the 1970s, satellite thermal infrared images from the US National Oceanic and Atmospheric Administration's Advanced Very

High-Resolution Radiometer indicated that the Gulf Stream was a single filament; additional research revealed a shifting, sinuous course with interannual variability.

Satellite studies of sea-surface temperature, in addition to exposing the basic structure of the Gulf Stream, led to the discovery of warm-core rings, which form between the North American continental shelf and the Gulf Stream due to the separation of a north-extending meander. The physical, chemical, and biological features of the centre of these eddies, which are 100-200 km in diameter, are frequently identical to those of the Sargasso Sea. When the interaction with the Gulf Stream is very vigorous, major changes in ring structure can occur in as little as 2 to 5 days. Following that, studies that combined satellite observations of sea-surface temperature, chlorophyll a, and sea-surface height with shipboard sampling revealed that cyclonic eddies have higher nutrient contents than anticyclonic eddies. Eddies, unlike Rossby waves, can forcefully upwell nutrient-rich waters, increasing phytoplankton productivity (Hoarfrost et al., 2019).

Vulnerability to Climate Change and Variability

Support for decisions in adjusting to trends and interannual variability in climate has been identified as a research need. Improving understanding of regional variability and the expected impacts of climate change could aid in identifying vulnerable people and civilizations. Science can help identify new alternatives and techniques for minimizing the extent of climate change or adapting to its effects, as well as improving existing ones. To do so, climate change indicators must be monitored, i.e., metrics important for understanding climate change and providing insight into environmental sustainability must be identified and tracked in categories such as the cryosphere, terrestrial ecosystems, hydrology and water resources, atmosphere, human health, oceans (both physical and biological/chemical), and natural disasters (Kogo et al., 2021).

Integration of knowledge about possible changes in the physical environment with measures of population and government resilience aids in identifying vulnerable people and societies. Demographic data, such as infant mortality; economic data, such as consumption and wealth distribution; sociological statistics, such as capitalism, globalization, and equality; and measures of democracy vs autocracy are examples of such indicators.

Data-Intensive Scientific Approach

Large volumes of data, the integration of diverse databases, and the associated computational hurdles are driving this trend. Many environmental challenges would be intractable without the use of data science due to the volume and/or complexity of the data involved. The following sections discuss some of the fourth paradigm features of environmental science

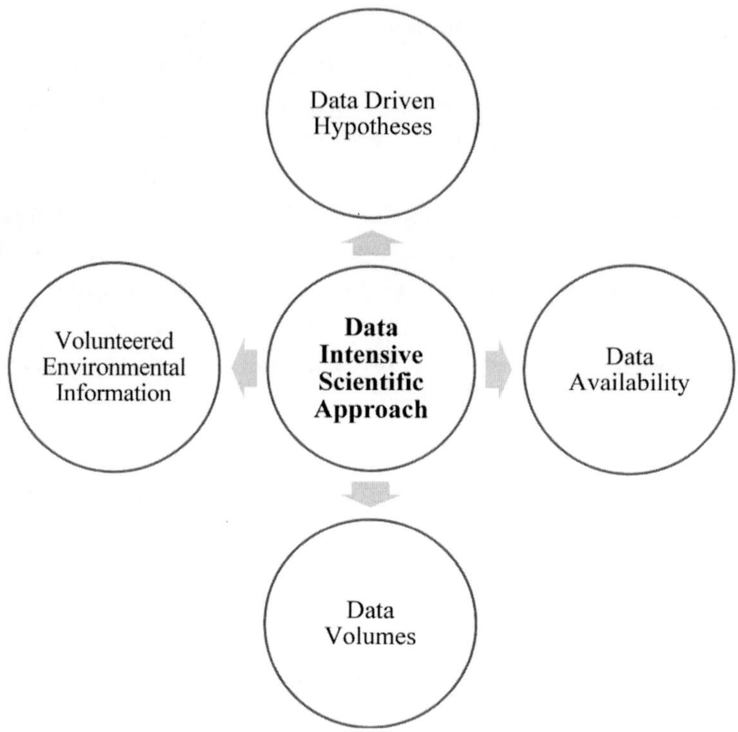

Figure 3. *Data Intensive Scientific Approach in Environmental sciences*

Figure 3 shows the different approach for data intensive scientific for environmental science that includes data volumes, Volunteered Environmental Information, Data Driven Hypotheses and Data Availability etc. These approaches of data driven approaches have been discussed in detailed below.

Data-Driven Hypotheses

Science evolves through hypothesis-driven investigation yet developing good hypotheses has always been an art and a trade, relying mostly on the inspirations of experienced researchers. By developing potential hypotheses from the correlations uncovered by mining environmental datasets, data science applied to the ever-increasing amounts of environmental information can help automate this process. Although some prominent scientists have expressed reservations about this viewpoint, framing the debate in terms of hypotheses versus informatics overlooks an important point: informatics can greatly aid traditional hypothesis-driven research because data-mining techniques can uncover significant large-scale correlations that provide rich information (Xie et al., 2020).

Some data are acquired deliberately to test a theory, but many of our data in the environmental disciplines are collected for operational purposes. Weather stations and river flow gauges, for example, were not intended to evaluate whether climate is changing and how those changes affect runoff, but they now give long-term data that helps answer such concerns. Informatics tools are critical in such investigations because they allow for the search for unexpected correlations and aberrant patterns. The growing trend of atmospheric CO2, the Antarctic ozone hole, the role of polar stratospheric clouds, the structure of the Gulf Stream, and the presence of warm-core rings are all examples.

Scientists use data visualization to explore hypotheses and present findings. Data exploration is an intrinsically creative activity that necessitates the discovery and visualization of relevant data, the discovery of candidate links, collaboration with peers, and the dissemination of discoveries. Historically, the difficulty was often that of data transformations and presentations, but today's data quantities, dataset numbers, and complexity of analyses necessitate the development of informatics tools, such as those for examining three- and four-dimensional data. In the perspective of Karl Popper's differentiation between the bucket and the searchlight as ways the mind learns (he preferred the searchlight, which allows for trial-and-error learning), informatics tools can help us learn through means other than duplicating the brain (Zhang et al., 2022).

Data Availability

Accessibility is supposed to be as simple as obtaining any other type of digital information. The remaining hurdles for data availability are structural in nature and policy based. After identifying a data source, the interfaces mentioned above allow it to be examined and exploited However, simply discovering the existence of a potential data source is still a substantial task. A potential technique for announcing data and service availability is emerging

It makes use of the Atom syndication format used by Web sites to give a dynamically updated table of contents in the form of a standard XML document. Portals can be developed as well-known places of reference, gathering dataset and service information from data providers.

Table 1. *A list of the different types of databases related to environmental informatics.*

S.No	Databases	Description
1	Agricultural & Environmental Science Database	This database covers journal articles and papers in environmental sciences from several disciplines.
2	Web of Science	This collection consists of nearly 18,000 periodicals.
3	Ecology Abstracts	Articles on evolutionary biology, economics, and systems analysis as they pertain to ecosystems, or the environment are indexed.
4	GreenFILE	Indexing, abstracts, and full-text articles on human influence on the environment.
5	CQ Researcher	Reports on political and social concerns, including frequent updates on health, foreign affairs, education, the environment, technology, and the US economy.
6	CAB Direct	4 million+ articles, books, reports, and proceedings in animal and plant sciences, human health, ecology and environmental sciences, agricultural economics and rural studies, and leisure and tourism are indexed.
7	BIOSIS Citation Index	A comprehensive index of life sciences and biomedical research that includes pre-clinical and experimental research, methodologies and instruments, animal studies, and much more.
8	PAIS International	Covers public affairs, public and social policy, and international relations, with some full-text articles available.
9	Water Resources Abstracts	Journal article abstracts, books, and other studies on water resources.
10	TOXLINE	An index to publications on various elements of toxicity. It is produced by the National Library of Medicine and covers the most recent five years of articles, books, papers, and reports.

Data Volumes

End-to-end input/output bandwidth is the limiting factor in computational environments. System architectures must seek to reduce the distance between processing and data. Software frameworks such as Hadoop can automate many of the distribution and coordination aspects. Map/reduce has the virtue of being able to scale up to take use of available resources. Database support for scientific data types, particularly arrays, is the second technology that will influence how environmental informatics works with very big datasets.

The key benefit of a DBMS for big data sets is its ability to manage the mapping between high-level queries and low-level storage operations. There are two potential answers to this problem: Rasdaman's raster data manager sits atop a traditional relational database or SciDB, an entirely new database system that supports multidimensional arrays (Baashar et al., 2020).

Volunteered Environmental Information

Volunteered information is a term that has emerged to describe information contributed to the Web by no specialists. Many of the Web's most prominent collaborative artifacts have elements that aid in the recovery of environmental data. Geotagging, the embedding of geographic position information in an otherwise no geospatial item, is a particularly important tool. Photo-sharing site Flickr may leverage latitude/longitude coordinates to build what is effectively a user-contributed border for a place, which may be especially valuable for regions with no official physical footprint (e.g., Southern California). The Open Street Map project is creating online maps of the world's metropolitan regions using active volunteers (Sadeghi-Niaraki et al., 2020).

Conclusion

The growing Web support for provenance should aid in deciding if a data product is appropriate and trustworthy. Digital environmental databases will require active management (curation) if they are to last longer than a few decades. The introduction of distributed announcement systems (data casting) and the proliferation of data regulations demanding sharing and free access could aid data availability. Two technologies, map/reduce and DBMSs that can appropriately describe and handle environmental information structures, are poised to assist meet the requirement to transfer processing to the data cloud. Non-scientists will be able to gather environmental information thanks to ubiquitous location sensing and geotagging, ushering in a new era of voluntary environmental data collection. information. Environmental informatics is concerned with environmental issues by employing data science approaches on huge multidimensional, complicated sets of spatiotemporally discrete and continuous data. Because environmental datasets have extensive information for accurate interpretation, they are best stored in complicated file formats or DBMSs. Geospatial Web APIs provides remote access as well as dynamic subsetting of environmental data. Servers and clients may be designed independently thanks to standardization. When permanent IDs are assigned to data packages, they become citable, greatly assisting the data publication process. In the present computer landscape, input/output bandwidth is the primary constraint on total system performance for huge datasets. This fosters a strategy of shifting processing to data rather than data retrieval.

Disclaimer

None

References

Afanasyev VY, Chernyshenko SV, Stylianakis SV. Biomonitoring information technologies as an important factor in providing a comfortable urban environment. *In Institute of Scientific Communications Conference* (2019) 2108:2120.

Appleyard SA, Maher S, Pogonoski JJ, Bent SJ, Chua XY, McGrath A. Assessing DNA for fish identifications from reference collections: the good, bad and ugly shed light on formalin fixation and sequencing approaches. *Journal of Fish Biology* (2021) 98(5):1421-32.

Baashar Y, Alhussian H, Patel A, Alkawsi G, Alzahrani AI, Alfarraj O, Hayder G. Customer relationship management systems (CRMS) in the healthcare environment: A systematic literature review. *Computer Standards & Interfaces* (2020) 71:103442.

Beretta V, Desconnets JC, Mougenot I, Arslan M, Barde J, Chaffard V. A user-centric metadata model to foster sharing and reuse of multidisciplinary datasets in environmental and life sciences. *Computers & Geosciences (*2021) 154:104807.

Chen J, Nurdin HI, Yamamoto N. Temporal information processing on noisy quantum computers. *Physical Review Applied* (2020) 14(2):024065.

Chowdhury KP. Supervised machine learning and heuristic algorithms for outlier detection in irregular spatiotemporal datasets. *Journal of Environmental Informatics* (2019) 33(1):1-6.

Cioruta B, Coman M. From enviromatics to sustainable informatics: beyond the definition and conceptual delimitations. *Ecological Safety and Balanced Use of Resources* (2019) 1(19):7-18.

Dhomse SS, Feng W, Montzka SA, Hossaini R, Keeble J, Pyle JA, Daniel JS, Chipperfield MP. Delay in recovery of the Antarctic ozone hole from unexpected CFC-11 emissions. *Nature communications* (2019) 10(1):1-2.

Dundore-Arias JP, Eloe-Fadrosh EA, Schriml LM, Beattie GA, Brennan FP, Busby PE, Calderon RB, Castle SC, Emerson JB, Everhart SE, Eversole K. Community-driven metadata standards for agricultural microbiome research. *Phytobiomes journal* (2020) 15;4(2):115-21.

Gan T, Tarboton DG, Horsburgh JS, Dash P, Idaszak R, Yi H. Collaborative sharing of multidimensional space-time data in a next generation hydrologic information system. *Environmental Modelling & Software* (2020) 129:104706.

Gangopadhyay A, Gawarkiewicz G, Silva EN, Monim M, Clark J. An observed regime shift in the formation of warm core rings from the Gulf Stream. *Scientific reports* (2019) 23;9(1):1-9.

Ghosh A, Maiti R. Development of new Ecological Susceptibility Index (ESI) for monitoring ecological risk of river corridor using F-AHP and AHP and its application on the Mayurakshi river of Eastern India. *Ecological Informatics* (2021) 63:101318.

Gordon S, Habermann T. The influence of community recommendations on metadata completeness. *Ecological informatics* (2018) 43:38-51.

Hao M, Liu G, Xie D. Hyperspectral face recognition with a spatial information fusion for local dynamic texture patterns and collaborative representation classifier. *IET Image Processing (*2021) 15(8):1617-28.

Hardisty AR, Michener WK, Agosti D, García EA, Bastin L, Belbin L, Bowser A, Buttigieg PL, Canhos DA, Egloff W, De Giovanni R. The Bari Manifesto: An interoperability framework for essential biodiversity variables. *Ecological informatics* (2019) 49:22-31.

Hoarfrost A, Balmonte JP, Ghobrial S, Ziervogel K, Bane J, Gawarkiewicz G, Arnosti C. Gulf Stream ring water intrusion on the Mid-Atlantic Bight continental shelf break affects microbially driven carbon cycling. *Frontiers in Marine Science* (2019) 11;6:394.

Hori T, Agata R, Ichimura T, Fujita K, Yamaguchi T, Iinuma T. High-fidelity elastic Green's functions for subduction zone models consistent with the global standard geodetic reference system. *Earth, Planets and Space* (2021) 73(1):1-2.

Imani M, Ghassemian H. An overview on spectral and spatial information fusion for hyperspectral image classification: Current trends and challenges. *Information fusion* (2020) 59:59-83.

Jalal A, Salman A, Mian A, Shortis M, Shafait F. Fish detection and species classification in underwater environments using deep learning with temporal information. *Ecological Informatics* (2020) 57:101088.

Jaspersen V, Fabritius M, Ahrens M, Huth J, Leinenkugel P, Kuenzer C. A Web-Based Environmental Information and Visualization System for Interdisciplinary Project Contexts: An Insight. *In Advances and New Trends in Environmental Informatics* (2018) 3-20.

Khaiter PA, Erechtchoukova MG. Conceptualizing an Environmental Software Modeling Framework for Sustainable Management Using UML. *Journal of Environmental Informatics (*2019) 34(2).

Kogo BK, Kumar L, Koech R. Climate change and variability in Kenya: a review of impacts on agriculture and food security. *Environment, Development and Sustainability* (2021) 23(1):23-43.

Martin O, Scholze M, Ermler S, McPhie J, Bopp SK, Kienzler A, Parissis N, Kortenkamp A. Ten years of research on synergisms and antagonisms in chemical mixtures: A systematic review and quantitative reappraisal of mixture studies. *Environment International (*2021) 146:106206.

Michener WK. Creating and Managing Metadata. *In Ecological Informatics* (2018) 71-88. Springer, Cham.

Nag R, Cummins E. Analysis of the levels of metal (loid) s in environmental compartments in Ireland towards a screening measure for potential relative risk using open-source datasets. *Journal of Environmental Management* (2021) 298:113531.

Nassour J, Leykin D, Elhadad M, Cohen O. Computational Text Analysis of A Scientific Resilience Management Corpus: Environmental Insights and Implications. *Journal of Environmental Informatics (*2020) 1;36(1).

Paul P, Aithal PS, Bhimali A, Kalishankar T. Environmental informatics vis-à-vis big data analytics: The geo-spatial & sustainable solutions. *International Journal of Applied Engineering and Management Letters (IJAEML) (*2020) 4(2):31-40.

Paul PK, Aithal PS. Environmental Informatics and its Features, Functions and Stakeholders: A Comprehensive Overview. *Educational Quest* (2020) 11(1):1-5.

Pipattanasomporn M, Chitalia G, Songsiri J, Aswakul C, Pora W, Suwankawin S, Audomvongseree K, Hoonchareon N. CU-BEMS, smart building electricity consumption and indoor environmental sensor datasets. *Scientific Data* (2020) 7(1):1-4.

Porter JH. Scientific databases for environmental research. *Ecological Informatics* (2018) 27-53.

Sadeghi-Niaraki A, Jelokhani-Niaraki M, Choi SM. A volunteered geographic information-based environmental decision support system for waste management and decision making. *Sustainability* (2020) 12(15):6012.

Spolaor A, Burgay F, Fernandez RP, Turetta C, Cuevas CA, Kim K, Kinnison DE, Lamarque JF, De Blasi F, Barbaro E, Corella JP. Antarctic ozone hole modifies iodine geochemistry on the Antarctic Plateau. *Nature communications* (2021) 12(1):1-9.

Tandule CR, Gogoi MM, Kotalo RG, Babu SS. On the net primary productivity over the Arabian Sea due to the reduction in mineral dust deposition. *Scientific reports* (2022) 12(1):1-2.

Wardani DW, Wahono EP. Spatio-temporal dynamics of tuberculosis clusters in Indonesia. *Indian Journal of Community Medicine: Official Publication of Indian Association of Preventive & Social Medicine* (2020) 45(1):43.

Wei HW, Hassan M, Che Y, Peng QK, Wang Q, Su YL, Xie B. Spatio-Temporal Characteristics and Source Apportionment of Water Pollutants in Upper Reaches of Maotiao River, Southwest of China, from 2003 to 2015. *Journal of Environmental Informatics* (2021) 37(2).

Wöllauer S, Zeuss D, Magdon P, Nauss T. RSDB: an easy to deploy open-source web platform for remote sensing raster and point cloud data management, exploration and processing. *Ecography* (2021) 44(3):414-26.

Xie J, Hu K, Zhu M, Guo Y. Data-driven analysis of global research trends in bioacoustics and ecoacoustics from 1991 to 2018. *Ecological Informatics* (2020) 57:101068.

Xu J, Wang W, Wang H, Guo J. Multi-model ensemble with rich spatial information for object detection. *Pattern Recognition* (2020) 99:107098.

Zhang J, Fu P, Meng F, Yang X, Xu J, Cui Y. Estimation algorithm for chlorophyll-a concentrations in water from hyperspectral images based on feature derivation and ensemble learning. *Ecological Informatics* (2022) 71:101783.

Zhang L, Guo H, Jiao H, Liu G, Shen G, Wu W. A polar coordinate system based on a projection surface for moon-based earth observation images. *Advances in Space Research* (2019) 64(11):2209-20.

Zhao M. Human spatial representation: what we cannot learn from the studies of rodent navigation. *Journal of neurophysiology* (2018) 120(5):2453-65.

Chapter 9

Methods and Challenges Regarding the Journey of Biomedical Waste to Zero Waste

Sirjan Kaur and Rachna Chaturvedi*
Amity Institute of Biotechnology, Amity University Uttar Pradesh Lucknow Campus

Abstract

Waste is a product that is no longer useful and has no intention to be utilized further. It is a carbon source of secondary nature that originates from humans as well as environmental activities. Human activities have high waste production and the waste produced is often tough to decompose. Biomedical waste is the waste originating from hospitals including blood samples discarded vaccines, and highly infectious microbial samples. Techniques such as incineration and landfills are often utilized to dispose of the waste generated but are a major cause of loss of carbon resources and depletion of fossil fuels. Medical waste can cause contamination and major health problems if not taken care of properly. It is necessary to not only accumulate the biomedical discard in a leakproof bag but also make sure that proper storage and transportation to sites are being done for proper treatment of the waste generated. The treatment of biomedical waste is done to make it non-infectious and safe for the environment. Different processes such as incineration, autoclaving, irradiation treatments, and pyrolysis are done to reduce the waste and make it manageable and nontoxic for the environment. These methods though effective cause secondary by-products and environmental pollution which includes the generation of a large amount of plastic waste, almost 85% of the total plastic production which is difficult to decompose. Several considerations need to be kept in mind to ensure proper management of waste and reduce the carbon footprint. The journey to zero waste requires the consumption and recovery of the products which is marked as waste without burning and releasing them into the environment. Biomedical waste to zero waste is a tedious yet achievable process and with few alterations can be acquired to generate the least amount of litter possible.

Keywords: Biomedical waste, Waste disposal, zero waste, Hazardous environment

Introduction

Biomedical waste also known as BMW is a rising concern during the past few years due to the covid-19 outbreak. The epidemic has caused a major rise in biomedical waste after a sudden requirement for masks, sanitizers, alcohol wipes, and other decontaminant products. The major requirement for vaccines has also caused a lot of waste accumulation in the form of syringes and plastic waste. These wastes cannot be reused without treatment due to their infectious nature and microbial contamination and are piled up requiring a lot of space and treatment. The covid-19 outbreak has not only resulted in medical waste accumulation but has also caused urgency for food shortage due to which people are hoarding perishable and many products unrequired by them which is not only causing waste products but also a need for more production of those items which are future waste. The packaging of these products is usually plastic which is the major contaminant and is non-biodegradable. BMW or biomedical waste consists of hazardous as well as non-hazardous waste, 85% of which is toxic or comes under hazardous waste. Methods that were used before the covid outbreak were designed for small-scale management and are failing due to abnormally large waste production after the outbreak, waste collection and waste management were not followed as stringently due to a lack of knowledge and scientific and economic resources. Apart from the danger of direct transmission of microbes from the waste, this waste if not treated properly accumulates in the water and soil causing contamination and adverse effects. The sharp needles and glassware can also cause physical injuries as well as direct transmission of a lethal disease. Diseases such as Covid-19, AIDS, Ebola, and other easily transmissible disease has led to the use of more disposable items usually made of plastic. Developing countries that don't have the facility or are often ignorant of the cleanliness around the hospital often dispose of the waste with the normal municipal waste which can be utilized by third parties by illegally utilizing the discarded waste in the black-market causing disease outbreaks. The compliance of the hospital bodies by proper education and staff for effective biomedical waste management is required and will be helpful in waste management. Hospital waste Treatment is an essential method that is the management of harmful waste and it requires particular experience and guidelines for safe disposal (Arshad N. et al. 2011, Chartier Y, et al. 2013)

The biomedical waste consists of-

- Human physical waste like tissues, organs, and body parts
- Creature squanders produced during research from veterinary clinics
- Microbial science and biotechnology squander

- Squander sharps like hypodermic needles, needles, surgical blades, and broken glass
- Disposed of prescriptions and cytotoxic medications
- Ruined waste, for example, dressing, swathes, mortar projects, material sullied with blood, cylinders, and catheters.
- Fluid waste from any of the tainted regions
- Burning debris and other compound squanders

Resilient Nature of BMW Management

In the situation of different pandemics, the rise in BMW production at the peak of the outbreak, as well as the clearance of contaminated not reusable masks and PPE are overpowering the current waste managing systems. Due to the high production of waste during outbreaks of different infectious diseases, from hospitals to houses, the waste management system is suffocating with more waste than it can be capable of handling. The expansion and overcrowding of hospitals have also made waste management a little less important. The segregation and storage of litter generated are ignored and the health workers are less trained in management skills. Also, the capability of the pathogen to continue active on surfaces for a long time would make us reconsider the rules recommending general segregation of the BMW (Doremalen et al., 2020).

The capacity constraints of in-situ incinerators and central treatment facilities result in the illegal dumping of wastes into suburban areas, streams, marshlands, etc., which raises public health concerns. The incinerators and machinery used to make the waste compact has been overflowing which is the cause of the illegal dumping of waste. To combat these issues, it is necessary to sterilize the general hospital waste which includes- syringes, and PPE kits and to set a strict code for proper disposal of waste in distributed dustbins to reduce the risk of contamination to the workers and cleaning staff. Sanitization and cleaning are major parts of the management to reduce the probability of spreading infection.

Certain methods apart from incineration which help in the management of biomedical waste include- Autoclaving, irradiation treatment, gas sterilization, thermal inactivation, and microwave heating.

In current times, new technology such as hhydrothermal carbonization is an emerging technology that is being used to assist in BMW management, this technique works on the principle of high pressure and temperature autoclaving which helps in treating non-infectious plastics. (Shen et al., 2017). The comprehensive harmony of HTC might be realized from further surveys (Sharma et al., 2020)

Biomedical waste management Globally

WHO describes the waste materials produced by different health and clinical organizations or research lab services throughout any medical and health actions as biomedical waste (BMW) (WHO, 2015). These types of medical and health actions incorporate identification as well as prophylactic, therapeutic, and mollifying therapies in the area of human as well as veterinary medicine. The waste materials generated in the course of health care commenced at home-based are also categorized as Biomedical waste (BMW). The BMW comprises wastes of sharp objects, contagious, clinical, therapeutic, compounds, and radioactive. It is made up of as non-hazardous waste (85%) and hazardous waste (15%) of which infectious and contagious waste is 10% and chemical or radioactive waste is 5% (WHO, 2017). Separately from the possibility of contact transmission, inappropriate discarding procedures of BMW can be the reason for unfavorable ecological impacts involving the contamination of soil and groundwater, destroying valuable microorganisms in septic structures, physical damages done with the sharps objects, etc., (Datta et al., 2018). Current knowledge from SARS-CoV, Ebola, and MERS-CoV infection occurrences focuses on the requirement for protected biomedical waste management for avoiding infections and their control.

The significance of BMW can also be connected to the disproportionate use of not reusable articles required because of the AIDS epidemic (Pandey et al., 2016; Rudraswamy et al., 2013). Surveys have stated that one of the most important causes of contamination is utilizing unintentional contact with discarded contagious kinds of stuff by health workers, staff, and consumers at the place of production. In advanced nations like the UK and the United States, hospital areas are lawfully guaranteed to avoid illnesses owing to negligence about the proper disposal of BMW (WHO, 2015). Nevertheless, the absence of satisfactory protection and carelessness in harmless management actions makes it a main concern for medical and health conveniences to safeguard BMW in opposition to illnesses connected to inadequate waste management procedures.

International Convention on BMW Management

Three known agreements which are involved are BMW management include- the Basel convention on hazardous waste, the Minamata convention on mercury, Stockholm convention on persistent organic pollutants which help in sustainable development and development of proper waste management policies.

Basel convention on hazardous waste includes 140 member countries whose objectives include safeguarding the environment and the human population against the effects of managing, disposing, and generating waste specifically from the healthcare department. Stockholm Convention is a worldwide treaty that helps manage the toxic chemicals called POPs which include dioxide and furans. These chemicals which are formed by the incineration of medical waste are released and accumulate in fat tissues of living organisms causing damage. It also provides best environmental practices (BEP) which help in reduction via recycling, recovery and proper collection of waste. (Chartier Y, et al. 2013) Minamata convention consists of 90 nations and is a global convention that protects the environment and humans from the adverse effects of mercury by slowly decreasing the production and usage of mercury-containing medical items such as thermometers. *Technical [Guidelines,31 October 2011.]*

Steps in BMW Management

Figure 1 provides the concept approximately following stages of participating in management of biomedical waste.

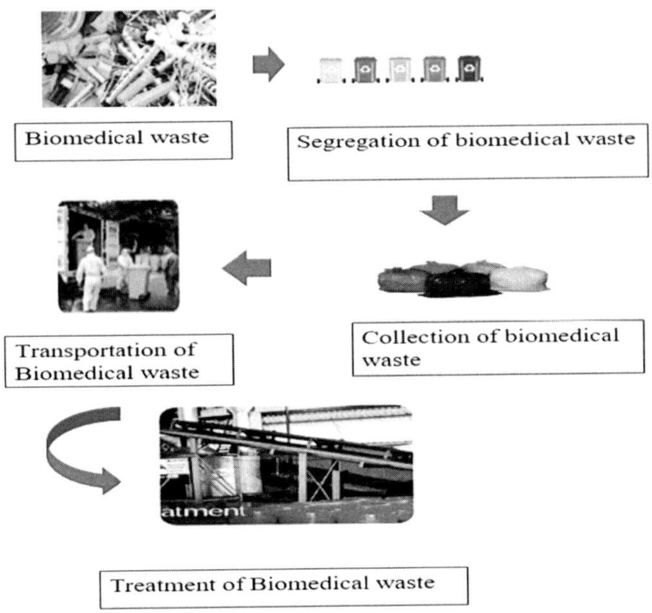

Figure 1. *Protocol for medical waste Treatment*

Segregation

Segregation is the process of separation of biomedical waste generated according to its nature. Waste can be of many types and segregation is essential cause different waste is treated according to its kind. Segregation is done at the point of generation of waste, the waste generated at hospitals or clinics is usually segregated by themselves, and measures are taken to never mix the different types of waste and cause contamination. The separation of waste is done by dividing it into four types- biodegradable waste, infected plastics, infectious waste, and glass sharps. Segregation of waste is made easier by allotting colors to different kinds of waste (Table 1 and Fig 2)

Table 1. *Segregation of waste distinguishing according to colour*

Yellow	This includes anatomical, soiled, and chemical waste. Discarded linen and chemotherapy-related waste are usually segregated in the yellow box.	Pre-treatment is done by disinfecting before handing the waste to the operator. The waste handed is then incinerated in a secondary combustion chamber.
Red	This waste includes contaminated plastic like syringes, plastic bottles, intravenous tubes etc.	Rubber waste and plastics are sterilized by autoclaving followed by shredding. Recyclable waste is disposed of with regular waste after treatment.
Blue	Metal implants and glassware such as broken Petri dishes, test tubes, and slides are added to the blue-colored dustbin.	No onsite treatment is required. The glass bottles are sterilized and disinfected followed by washing and then sent to recycle.
Grey/White	This waste consists of sharp things consisting of metals including- blades, broken scissors, discarded needles, and other surgical instruments.	Disinfection by high heat sterilization or autoclaving plus shredding and then disposed to concrete pit or sanitary landfill.
Black	Any other waste which does not fit the above categories is added to the black dustbin. It includes hazardous wastes such as pesticides, chemicals, disinfectants, batteries, etc.	Incineration and chemical treatment is the most used treatment method.
Green	Biological waste", is any category of organic waste that can be composted. It is most usually composed of refuse from gardens Grass clippings. Hedge trimmings.Branches.Plants.Flowers. Leaves.	Green waste is carried out to a particular composting location where it can be reprocessed.

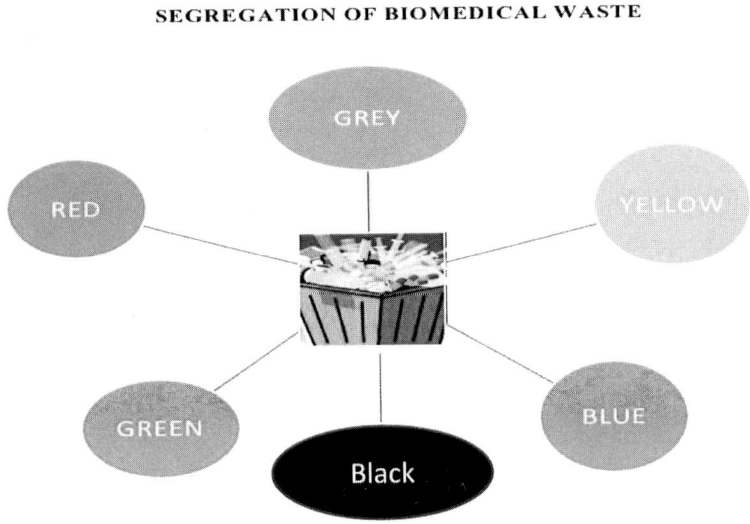

Figure 2. *Segregation of Biomedical waste*

Collection & Storage

Storage is done usually at the site of the generation of waste. Hospital wards and clinics are required to collect the waste in the process mentioned above and keep it within the department till the time the waste is picked up by the waste disposal operator. Collection centers plan a waste collection usually every 48 hrs. where they transport it to the treatment facility. The collection is the most essential step, placing bins at sites achieves 100% collection. Sharp materials are kept in leak-proof containers. Storage of the waste is done after collection in different bins. The color-coded bags are kept segregated and labeled to trace the waste to its source and a caution sign is placed showing the cytotoxicity and biohazards. Information on the labels such as pre-treatments or the hazardous nature of the waste is also mentioned so no contamination occurs

Transportation

The waste collected is transported every 48hrs and special containers such as trolleys or covered vans/ wheelbarrows are used to avoid spillage and accidental dropping of waste on the path. The open transporting vehicles are covered with a lid or tied before transportation. Before the transportation occurs, the bag containing BMW is documented and signed by the authority responsible for that place such as a doctor mentioning the date, quality, and destination. It is necessary to ensure that there is no contact or access to hazardous waste during transportation to any public. The driver should be trained and follow a given protocol in case any accidental spillage occurs. Personal safety devices such as gloves, masks, and rubber boots should be worn to handle the waste, usage of brooms, mops should be done to clean the vehicles before the loading of the waste.

Treatment and Disposal

Treatment is done to disinfect and decontaminate the waste or to reduce the volume of the waste for it to be compact before its disposal. Treatment of the waste is done usually according to the type of waste materials, but the two broad categories of its treatment are-

- **Burn Technology**- this includes open burning of the waste by using incinerators and kilns.
- **Non-Burn Technology**- this method includes the usage of chemical disinfection, microwave irradiation, autoclaving, deep burial, sanitary landfill, and encapsulation.

  a) **Autoclaving**- it is a thermal disinfection method used for reusable medical equipment which inactivates all the microorganisms and their spores. The temperature used is 121 degrees C at 15psi temperature for 15mins. It uses temperature and pressure to kill microbial organisms. (Fig 3)

  b) **Chemical Disinfection**- a method used for liquid waste like blood samples, sewage, and urine. The chemicals are added to the liquids to make them inert and kill microbial growth. (Fig 3)

  c) **Incineration** - it is the most used method of treatment to make the waste compact and is usually used for most hazardous health care waste which cannot be left untreated. This process reduces organic combustible waste into inorganic waste which is non-combustible with a reduction in its volume and weight. Types of incinerators used are- Single chamber incinerators, drum/ brick incinerators, pyrolytic incinerators, and rotatory kilns. (Fig 3)

  d) **Microwave Irradiation**- microwaves with a frequency of 2450 MHz and wavelength of 12.24 nm are used to destroy microbial actions. (Fig 3)

  e) **Land Disposal**- this step is usually followed by the treatment of waste by the above methods and the final product is taken to land on an open dump or in a sanitary landfill. It is necessary to recycle as much as possible to reach zero waste generation, but the unrecyclable products are dumped or are used for inertization in which the waste is cemented to be used in the cement industry. (Fig 3)

a) Autoclaving

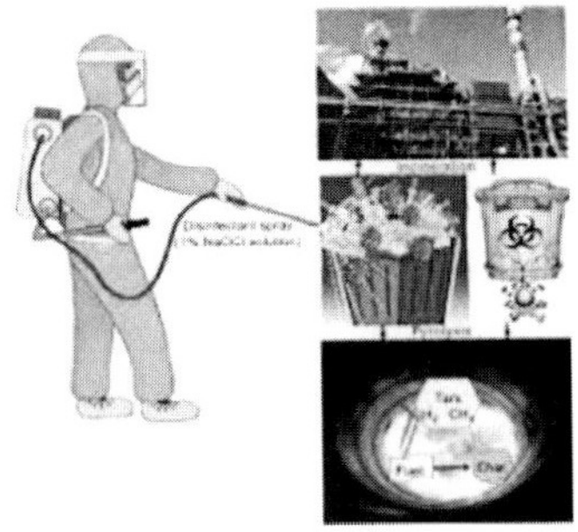

b. Chemical Disinfection

c) Incineration

d) Microwave radiation

e) Landfill disposal

Figure 3. *Medical Waste Treatment Methods*

BMW Rules In India

Biomedical waste management rules were renewed in 2016 and some of the features include as following-

- Slowly discard the use of chlorinated plastic bags, blood bags, and other plastic materials in two years.
- Disinfection or sterilization of waste generated in laboratories by pre-treatment methods.
- Major incidents reporting.
- Regular training for health care staff on the treatment of waste and its proper disposal by common biomedical and waste disposal facility (CBMWTF) and establishment of a global positioning system and barcoding for the handling of biomedical waste within a time limit of 1 year by the same (*Ministry of Environment,* 2016)

- Limiting the inclusion of furans and dioxins.
- Proper timely disposal and pickup of waste to the facility from the hospitals and clinics.
- Providing lands for the disposal of medical waste from the state government.
- Separation of liquid waste by pre-treatment before combining it with other waste materials.
- Maintenance of disposal registers with daily logs and monthly updates.
- No establishment of BMW facility along with the hospital care facility if the CBMWTF is present in seventy-five kilometers.
- Color coding of BMW based on waste treatment options to ensure the safety and proper waste treatment and avoid human health and environmental contamination. (Ministry of Environment, 2016)

Advantages of Implemented Rules

Implementation of rules established by the Indian government is strict and elaborate and brings changes in the way BMW is managed in India. The segregation procedure which was conducted before is reduced to four to make it easier for the health care workers to overcome the confusion that can be created by a large-scale category. The color coding of the disposed waste is beneficial for the separation and to treat the waste separately according to specific requirements. It has also been mandated that the pretreatment of laboratory-related waste is necessary. The policies established are environmentally friendly and are a step to make biomedical management easier and to reach the zero-waste mark by recycling. Reusing the waste is a crucial step that has been investigated while making this policy which states the phasing out of plastics in two years, this not only controls the generation of waste but also helps in decreasing the emission of dioxins and furans which are released during the burning process.

The incinerators used to burn the waste to reach compactness have been standardized to the particular emissions and secondary chambers are reduced to $50mg/nm^3$

Which will reduce the production of dioxide and furans which are released at a temperature greater than 600degree C.

The process of authorization, stringent procedures careful review of every step of the process are considered by the health sectors which is beneficial in implementing the new rules provided.

Technologies Crucial for the Journey of BMW To Zero Waste

The technologies which have been utilized not only help in making the waste compact but also are a step forward towards zero waste which conserves all resources and helps in recovery of all the waste so there is no need to dispose of it. The technologies used are divided into four categories and are called- chemical processes, biological processes, thermal processes, and irradiative processes.

The thermal process requires heat and is divided into low, medium, and high treatments.

- Low heat technology works at a temperature of 93degree C and involves autoclaves and microwaves. Autoclaves provide moist heat sterilization; steam is generated under pressure to decontaminate the waste. Autoclave first works to remove air from the chamber so that no cold spots or air pockets for proper sterilization, the steam is then passed with a saturation of 5% to create the proper sterilizing environment. Then the last step is followed by drying the chamber. Pre-vacuum/ high vacuum technologies are used for proper air evacuation. The evacuated air is also filtered using a HEPA filter to avoid contamination. Microwaves use heat and steam to disinfect. Materials such as radiological waste and organic compounds are not for microwave sterilization. Microwaves are efficient because of their minimal emissions during biomedical waste disposal but are usually costly to set up and release odor.
- Medium heat technology has a temperature of about 177 degrees C to 540 degrees C involving thermal depolymerization and reverses polymerization. Thermal depolymerization is the process of changing a compound to its monomer with the help of high heat. The final products are limited range, and the volatile compounds are burnt to form synthetic fuels. Reverse polymerization works on organic compounds and the application of high-energy microwaves for heat to break down the materials into simpler compounds. Waste soaks up the energy of the microwave which roots the decomposition to the molecular level as well as nitrogen level present helps in non-combustible reactions and the resulting waste is shredded.
- High heat technologies work at a very high temperature of 540 degrees C to 8300-degree C. Pyrolysis is the most significant procedure utilized which causes the carbon-based type of solid and liquid waste categories to evaporate at elevated temperatures in a pyrolysis chamber. Types of pyrolysis include- plasma pyrolysis, oxidation pyrolysis, induction-based pyrolysis, etc. the inert ash and fragments left after the initial treatment in the chamber are further treated in the second step where incineration of the fumes taking place at

1093 degrees C and the vapor is released. Plasma pyrolysis utilizes plasma lights to create plasma power. In plasma direct, the ionized gas can channel electric flow, yet because of its high opposition, the electrical energy is changed over completely to warm energy. The buildup produced incorporates carbon dark, vitrified glass totals, and metallic deposits. A wide assortment of waste can be obliterated in the innovation that is plasma-based, irresistible waste, sharps, plastics, dialysis squander, unsafe waste, chemotherapeutical waste, waste related to chemotherapy, and small-level radioactive waste (one omission is mercury which plasma agendas don't deal with). Many benefits of this framework are low emanation rate, squander buildup is latent and sterile, i.e., climate well disposed, and there is a decrease in volume (95%) and mass (80%-90%). Its drawback incorporates high capital expense, high activity cost, high electrical use, and the restricted life expectancy of plasma light which may create dioxins in the inadequately planned arrangement.

Numerous substances utilized for BMW removal are at present being worked on. The sort of waste treated by synthetic-based innovation incorporates societies, sharps, fluid waste, human waste, research facility squander, and delicate waste (dressing, wraps, and outfits). Unstable, semi-unpredictable natural mixtures, mercury, and radiological type of waste materials ought is not treated with compound-based strategies. The synthetic-based innovation requires a shut framework or is worked under regrettable tension and the depleted air must be gone through HEPA to defend against spray development during destruction. In substance-based innovation, destroying BMW is a must. The benefits of synthetic-based innovation incorporate a completely mechanized strategy, simplicity to utilize, simplicity of release of fluid emanating into the sewage, and no results of burning framed. The inconveniences incorporate poisonous side effects because of the enormous scope of chlorine and hypochlorite use, synthetic dangers, and the frequent creation of hostile smells. The synthetic-based innovation can be isolated into chlorine-and nonchlorine-based frameworks.

The chlorine-based framework utilizes sodium hypochlorite or chlorine dioxide. Sodium hypochlorite was quite possibly the earliest substance sanitizer used to treat BMW. Recently, it has been revealed that poisons like dioxins, radiance acidic corrosive, and chlorinated sweet-smelling compounds are delivered where sodium hypochlorite is utilized. Chlorine dioxide is shaky, so it is created and utilized nearby. It is the area of strength for a. It decays to frame salt, fewer poisons are delivered with its utilization, and it doesn't respond with ammonium or liquor.

Nonchlorine-based advancements use either gas, fluid, or dry substance to treat BMW. Large numbers of such frameworks have come into the market, and few are talked about here. The Steric Eco cycle 104 purposes a convenient chamber for gathering waste into which a peracetic corrosive-based purifier vial is added. After 10-12 min, peracetic corrosive cleans the waste, and sprayed microorganisms are kept from getting away by utilization of HEPA channel. The fluid profluent is released into the sewer and the waste goes as customary rash. (Emmanuel J. Washington, DC, 2001).

Squander decrease innovation utilizes antacid hydrolysis at high temperatures to change over human and microbial waste into an unbiased watery arrangement. It is utilized for human/tissue squandering, body liquids, and degradable packs. Likewise, it can deal with chemotherapy squandering. The salt additionally annihilates fixative sin tissues and different risky synthetic compounds, for example, glutaraldehyde and formaldehyde. This robotized framework has a steam-jacketed hardened steel compartment with a retainer for biodegradable waste like bone and teeth.

Lynntech's other innovation involves ozone for the sterilization of BMW. Being areas of strength for ozone obliterates organisms and converts them into sub-atomic oxygen. (Emmanuel J. Washington, DC 2001).

Ionizing radiation makes harms DNA and creating free revolutionaries brings about additional harmful proteins and catalysts of irresistible particles. The electron bar innovation produces ionizing radiation as a light emission energy electron impelled at fast to strike the objective. The loss to be dealt with is irresistible waste including human waste, research center waste, delicate waste (cloth, gauze, and outfits), and sharps. Unstable, semi-unpredictable natural mixtures, mercury, and radiological type of waste should be not treated in electron shaft units. The different benefits of this strategy are that it doesn't create poisonous emanation, no fluid profluent, no ionizing radiation after the machine is turned off, completely mechanized, and low functional expense. The weakness incorporates assurance from radiation openness, substantial safeguard a few feet thick around the framework, evacuation of ozone gas, and there is no reduction in squander volume and need for shredders or grinders.

Natural strategies for removal of BMW incorporate an arising framework called "Bio-converter" 9 Biomedical Disposal, Inc.). It utilizes an answer of the compound to clean clinical waste, and the subsequent slop is put through an extruder used to eliminate water for sewage removal and the strong waste is shipped off landfill. One more strategy for natural BMW removal is the utilization of biodegradable plastics. Numerous biomedical inserts worked with biodegradable plastics go through organic debasement with microbial extracellular proteins. These microorganisms use these biodegradable polymers as substrates under starvation and in the inaccessibility of reasonable substrate. Further examination should be finished for the enormous scope of financial assembling of biodegradable plastics.

Challenges during Journey of Biomedical Waste To Zero Waste in Covid

It is evident from data collected from CPCB that the COVID-19 waste Was in September 2020 (183 tonnes/day,). But in May 2021 it was recorded as 203 tonnes/ day. Consequently, the second wave of the Covid 19 indicates more infectious in comparison to the first wave (Singh 2021). The usage of all protective articles in all the facilities generates a load of biomedical waste While applying biomedical waste as a result of COVID-19 is combined with municipal waste and discarded in landfills may be the cause of ecological contamination and ultimately health threats (Gupta 2021). India is currently also struggling with appropriate BMW management and the establishment of the recent establishment of newer techniques (Goswami et al. 2021) The quick rise in BMW in a short period creates a general task in overcoming the condition. In a Covid 19 condition, where everybody is thinking about remedies regarding vaccination, the problem of biomedical waste also appears stuck between vaccination and disease avoidance (Das et al. 2021).

At present we are facing so many of the challenges for proper disposal of biomedical waste ultimately dealing with zero waste. There is no proper segregation of Biomedical Waste, following disposal of Surgical infected masks with other materials resulting in accumulation of hazardous waste. All types of waste that should be disposed in yellow bin, most of the time not following rules and regulations. At last, most important challenges are insufficient funds for implementation of rules and regulations.

Conclusion

In the medical services field, a large part of the focal point of examination is the improvement of new advances, procedures, drugs, drug conveyance frameworks, and so forth. It means quite a bit to ponder how to deal with the tremendous heaps of medical care waste that are and will get produced and their eco-accommodating administration. A forward leap is another medical care waste that is a need of an hour. Management of biomedical waste to zero waste is a long road and ought to be a common collaboration with committed government backing, great BMW rehearses followed by both medical services laborers and HCFs, constant observing of BMW rehearses, areas of strength for and. The use of the 3Rs reducing, reusing, and recycling is a crucial step towards zero waste motive and can be accomplished by small steps such as recycling packaging, reusing the equipment after autoclaving and sterilization, and recycling the materials for no waste disposal. Disposal of waste should be the last option for a safe environment and less waste generation. Living in a perfect and safe environment is our principal right. The mainstay of BMWM is the isolation of waste at source and WR. The ongoing BMWM 2016 principles are an enhancement over prior rules regarding further developed isolation, transport of waste, and removal strategies, to diminish natural contamination and guarantee the security of the staff, patients, and public. In addition, more utilization of non-PVC clinical gadgets and advancement of the fresher novel, eco-accommodating frameworks for removal of BMW ought to be energized. All members of BMWM ought to vow to ensure a cleaner and greener climate. As per the pandemic COVID-19 India, the CPCB of the Ministry of Environment,

Forests and Climate has announced regulations on the removal of waste produced for the duration of the treatment/diagnosis/quarantine of COVID-19 patients' will also check the regulations under the 2016 Biomedical Waste Management Regulations (Bio-Medical Waste Management (Amendment) Rules 2018) regularly so that COVID-19 and other waste produced by contagious diseases, will be gathered following protocol for extreme attention and transported to the "Biomedical Waste Disposal and Disposal Facility"

Disclaimer

None

References

Arshad N, Nayyar S, Amin F, Mahmood KT. Hospital waste disposal: A review article. Journal of Pharmaceutical Science & Research 2011;3(8):1412-19.

Chartier Y, et al. Safe management of wastes from healthcare activities. 2nd edition. Geneva: World Health Organization; 2013.

Das AK, Islam M N, Billah MM, Sarkar A COVID-19 pandemic and healthcare solid waste management strategy: a mini-review. Sci Total Environ 778:146220 (2021) PMID: 33711590, PMCID: PMC7932852, DOI: 10.1016/j.scitotenv.2021.146220

Datta P., Mohi G.K., Chander J. Biomedical waste management in India: a critical appraisal. J Lab Physicians. (2018) 10(1):6–14. [PMCfree article] [PubMed] [Google Scholar]

Doremalen V.N., Bushmaker T., Morris D.H., Holbrook M.G., Gamble A., Williamson B.N., Lloyd-Smith J.O. Aerosol and surface stability of SARS-CoV-2 as compared with SARS-CoV-1. N. Eng. J. Med. 2020;382(16):1564–1567.

Emmanuel J. Washington, DC: Health Care Without Harm;. Non-Incineration Medical Waste Treatment Technologies. Environ 778:146220.(2001)

Goswami M, Goswami PJ, Nautiyal S, Satya Prakash, Challenges, and actions to the environmental management of Bio-Medical Waste during COVID-19 pandemic in India. (2021) Heliyon 7(3): e06313.

Gupta G Management of COVID-19 waste. COVID-19 in the Environment Impact, Concerns, and Management of Coronavirus (2022), Pages 277-294

Pandey A., Ahuja S., Madan M., Asthana A.K. Bio-medical waste management in a tertiary care hospital: an overview. J. Clin. Diagn.Res. (2016);10(11) DC01.

Rudraswamy S., Sampath N., Doggalli N. Global scenario of hospital waste management. Int. J. Environ. Biol. (2013);3(3):143–146.

Sharma H.B., Sarmah A.K., Dubey B. Hydrothermal carbonization of renewable waste biomass for solid biofuel production: discussion on process mechanism, the influence of process parameters, environmental performance and fuel properties of hydro char. Renew. Sustain. Energy Rev. (2020 123.

Shen Y., Yu S., Ge S, Chen X, Ge X, Chen M., Hydrothermal carbonization of medical wastes lignocellulosic biomass for solid fuel production from lab-scale to pilot-scale. Energy. (2017) ;118:312–323.

Singh S COVID-19 will place India's biomedical waste management under terrible strain (downtoearth.org.in) (2021)

Technical guidelines for the environmentally sound management of used and waste pneumatic tyres, revised final version, 31 October 2011

Technical guidelines for the environmentally sound management of used and waste pneumatic tyres, revised final version, 31 October 2011

Ministry of Environment, Forest and Climate Change, Government of India. The Bio-Medical Waste Management Rules, 2016, Notification, New Delhi, 28th March 2016.

WHO. Safe management of wastes from health-care activities. The World Health Organization. (2015).

WHO. Report on health-care waste management (HCWM) status in Countries of the South-East Asia Region (SEA-EH-593). The World Health Organization. (2017).